Biomimetic and Bioorganic Chemistry III

Chemistry of
Multi-Armed Organic Compounds
F. M. Menger

Calculation of Interaction Energies in
Host-Guest Systems
M. Badertscher, M. Welti, P. Portmann and E. Pretsch

Design of Biospecific Compounds which
Simulate Enzyme-Substrate Interaction
K. Tanizawa and Y. Kanacka

Recent Developments in the Field of
Biologically Active Peptides
G. Schmidt

AKADEMIE-VERLAG BERLIN

Topics in Current Chemistry · Vol. 136

Biomimetic and Bioorganic Chemistry III

Biomimetic and Bioorganic Chemistry III

With 52 Figures and 13 Tables

AKADEMIE-VERLAG BERLIN
1987

Die Originalausgabe erscheint im Springer-Verlag
Berlin · Heidelberg · New York · London · Paris · Tokyo
in der Schriftenreihe „Topics in Current Chemistry“,
Volume 136

Vertrieb ausschließlich für die DDR und die sozialistischen Länder
Akademie-Verlag Berlin

ISBN 3-540-16724-2 Springer-Verlag Berlin · Heidelberg · New York · Tokyo
ISBN 0-387-16724-2 Springer-Verlag New York · Heidelberg · Berlin · Tokyo

ISBN 3-05-500285-7 Akademie-Verlag Berlin

Erschienen im Akademie-Verlag Berlin, DDR-1086 Berlin, Leipziger Straße 3—4
Lizenznummer: 202 · 100/468/87
Printed in the German Democratic Republic
Gesamtherstellung: VEB Druckerei „Thomas Müntzer“, 5820 Bad Langensalza
LSV 1275
Bestellnummer: 763 703 3 (3072/136)
11200

Editorial Board

Table of Contents

Table of Contents of Volume 128

Table of Contents of Volume 132

Chemistry of Multi-Armed Organic Compounds

Fredric M. Menger

Department of Chemistry, Emory University, Atlanta, Georgia 30322/U.S.A.

Table of Contents

1 Introduction

When chemists carry out a reaction in the laboratory, they hope that a suitable fraction of the molecular collisions will possess sufficient energy and the proper geometry for a successful transformation. If the reaction rate is slow, little can be done about increasing the *percentage* of successful hits. The best one can do is to increase the temperature and thereby increase the *total number* of collisions per unit time. This is a fairly crude procedure and not practical for biological systems which must operate within a well-defined and narrow temperature range. Thus, enzymes do not depend on highly inefficient random collisions, but function instead by collecting the necessary species in a "cavity" or "pocket" located on the protein surface. The correct orientation and distance, forced upon the reactive functionalities within this restricted volume, greatly accelerates reaction rates. Clearly, organic chemists must synthesize cavity-bearing molecules in order to satisfactorily model enzymes. This paper deals with one particular type of "space-encompassing" compound: the multi-armed system.

Our own interest in multi-armed compounds stems from work with micelles[1]. Micelles are spherical aggregates of roughly 50–100 surfactant molecules each comprised of a polar head group and a long hydrocarbon chain. A huge number of papers have been devoted to micellar reactions owing, in part, to their superficial resemblance to enzyme-catalyzed processes. Micelles bind compounds, for example, with association constants rivaling those for many enzymes and their substrates. Binding to micelles often leads to catalyzed reactions obeying Michaelis-Menten kinetics [2]. Stereoselectivity is possible with micelles composed of chiral surfactants [3]. Yet there is a rather serious disadvantage of micellar systems: to observe micellar effects, the surfactant concentration must exceed a critical micelle concentration. Otherwise, the surfactant exists entirely in the monomeric state. Thus, it is natural to think of tying the chains together by covalent linkages in order to prevent the chains from dissociating from each other. Such a multi-chain or "multi-armed" compound could behave like a micelle at *all* concentrations.

The above paragraph reflects our original motivation for studying multi-armed species. Other investigators in the area undoubtedly had different reasons for constructing molecules with many appendages. Probably the likelihood of interesting and perhaps unique properties stimulated much of the research. And in all cases, I feel certain, the investigators were intrigued by the admittedly anthropormorphic resemblance of their systems to the human hand and its capability to grasp.

Although this review is not exhaustive, the examples herein should serve to illustrate the important properties of multi-armed compounds as far as they are now known. A concise, almost compressed, format should provide the reader an overview with a minimum amount of reading time.

2 Multi-armed Polyethers

In 1974 Vögtle and Weber published a paper entitled "Octapus Molecules" [4]. The paper describes a hexasubstituted benzene derivative (1*a*) which shows "remarkable phenomenological parallelisms to the mode of food capture by an octopus using its suction pads." Compound (*1a*), a water-insoluble oil, was found to be a powerful ligand for cations. Thus, a dichloromethane solution of (*1a*) is able to extract metal picrates (Li^+, Na^+, K^+, Ca^{2+}, Mg^{2+}, etc.) from an aqueous phase (and do so more rapidly and completely than do the corresponding solutions of crown ethers). Neither (*1a*) nor crown ethers are able to extract heavy-metal salts ($CuCl_2$, $NiCl_2$, $Ce(SO_4)_2$, etc.) from water.

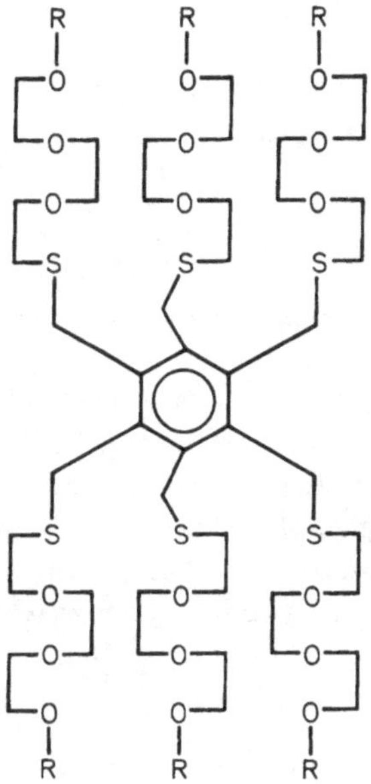

1a

The complexing ability of the multi-armed polyethers diminishes when (a) the arms are shortened so as to possess only two oxygens per arm and (b) the number of arms are successively reduced. This is illustrated with formula (*1*)–(*4*) (*a*) and (*b*).

Apparently, the most stable complex forms when six chains residing on the *same* side of the benzene ring surround the metal ion. Three chains on the same side of the aromatic ring can best enclose a "cavity" favourable to metal ion complexation if the chains are in a 1,3,5 relationship [cf. (*2*)] as opposed to 1,2,3 [cf. (*3*)] or 1,2,4. Ligand properties for the 2-chain molecules are found only with the 1,2-isomer (*4*).

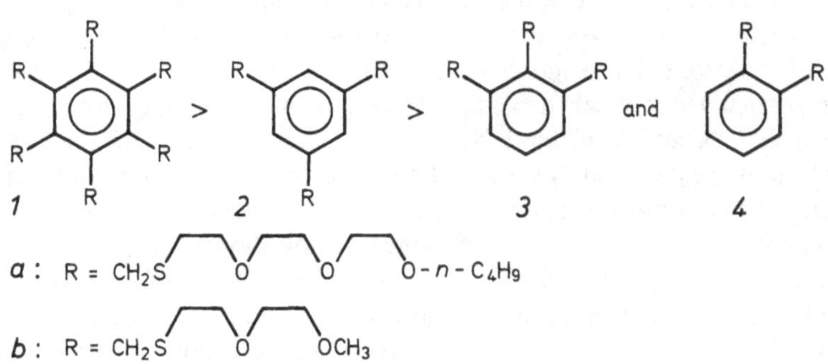

Steric overcrowding in (*1a*) is apparent from a broad α-methylene proton signal. Broadening, indicative of hindered internal rotation, varies with the nature of the arms: shorter arms impede rotation less than long ones

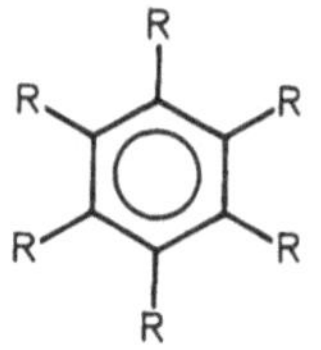

R	Signal broadening (α-CH_2)
n-$C_{12}H_{25}S$—	Marked
n-C_4H_9S—	Moderate
CH_3S—	None

Fig. 1. ^{1}4 nmr characterization of (1*a*)

(*Fig. 1*). Reducing the number of arms also decreases the crowing and hence the signal line-widths.

Vögtle and co-workers [5] synthesized a host of neutral "tripod" and "tetrapod" ligands (*6*). Cation selectivity depends on the particular end-group R. When the end-group is a quinolyloxy unit, the ligand binds Ba^{2+} with an association constant close to 10^3 M^{-1} (at least an order of magnitude higher than that for the corresponding "dipod" ligand). When phthalimido is the end-group, one can form and isolate a solid complex with $FeCl_3$. Various other end-groups give solid complexes with $ZnCl_2$, RbI, and $Th(NO_3)_4$.

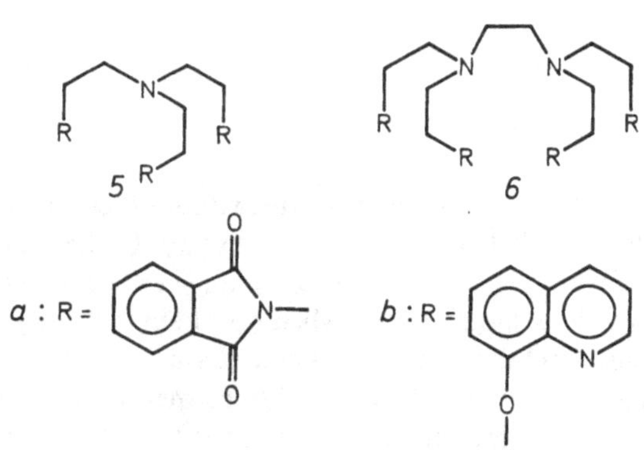

Hyatt [6] attached six polyether chains onto a rigid and bowl-shaped cyclotriveratrylene framework (*7*). These chains are capable of surrounding a space which incorporates metal ions. The length of the polyether arms, $-(C_2H_4O-)_nR$, does not appear to be particularly critical to binding as long as n > 1. For example, the derivatives with n = 2, 3, and 4 all bind Na^+ rapidly and strongly, and they all bind Mg^{2+} slowly and weakly. On the other hand, the stereochemistry and conformational rigidity of the supporting framework does play a crucial role in binding. No complex is observed, for example, with an 8-armed analog shown with formula (*8*). Examination of molecular models show that conformationally mobile aromatic rings do not readily adopt a bowl-shaped configuration necessary for the chains to enclose a cavity. As a consequence, crown ether-like behavior is not observed.

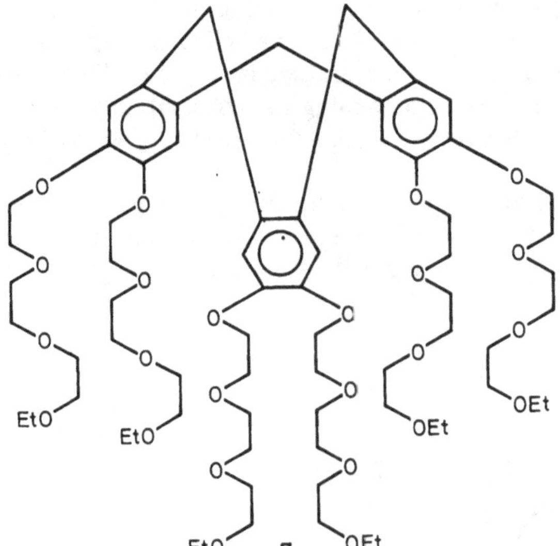

7

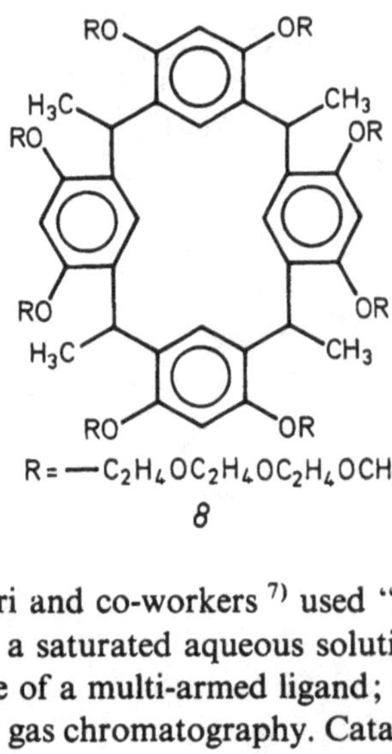

$R = —C_2H_4OC_2H_4OC_2H_4OCH_3$

8

Montanari and co-workers [7] used "polypode" ligands (*9*) as phase-transfer catalysts. Thus, a saturated aqueous solution of KI was stirred with *n*-octyl bromide in the presence of a multi-armed ligand; the resulting production of *n*-octyl iodide was followed by gas chromatography. Catalyst activity was found to vary with the hydrophobicity of the arms. When the chains are terminated by butyl groups, 72 hours

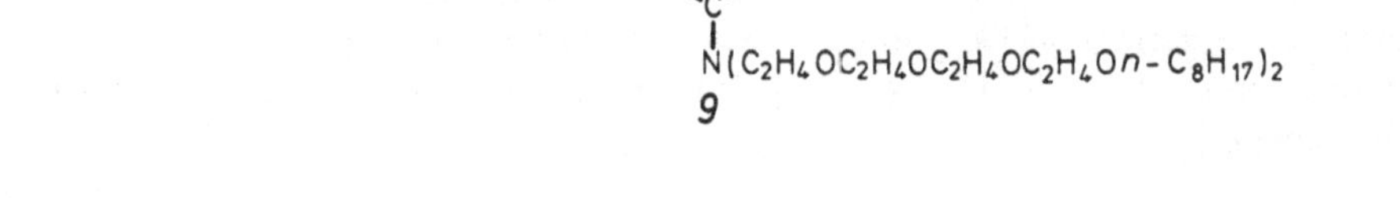

9

are required for an 82% yield. When the chains are terminated by octyl groups, only 3 hours are needed for an 85% yield. The latter efficiency approximates that of hexdecyltrimethylphosphonium bromide and certain cryptates. The authors speculate that an associative apolar interaction among the terminal alkyl groups leads to a relatively stable cavity within the polyoxymethylene chains which favors chelation of metal ions.

Although the majority of compounds discussed in this paper have linear arms, cases are known in which a central unit bears multiple ring substituents. A good example (*10*) was investigated by Weber [8]. This material is highly soluble in water

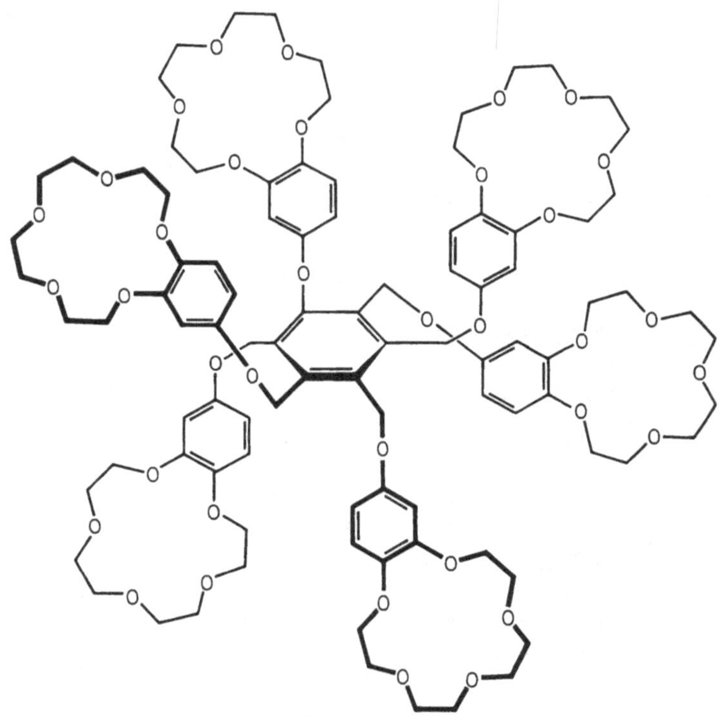

10

but relatively insoluble in nonpolar solvents such as chloroform. Interestingly, the properties of the compound resemble those of a surfactant:

(a) Aqueous solutions of (*10*) foam;

(b) Concentrating (*10*) in water elevates the viscosity until there is formed a reversible, opaque, and almost immobile gel;

(c) Warming aqueous solutions of (*10*) causes clouding at 52–54 °C;

(d) Light scattering experiments indicate the formation of micelles with a critical micelle concentration of 1.3×10^{-4} M and an aggregation number of 50. Weber states that the surfactant-like behavior was unexpected, and the present author is likewise astonished. Undoubtedly, many other pleasant surprises lie in store for the octapus chemist.

3 Multi-armed Compounds with Ion-Terminated Chains

"Tentacle" molecules having ion-terminated hydrocarbon chains that radiate from a central unit are relatively rare in the literature. Suckling [9] examined benzene-1,3,5-tricarboxylic acid esterified with three $[CH_2]_{11}NR_3^+$ groups. The resulting tentacle molecule (*11*) forms complexes with small aromatics in acetonitrile but not in methanol.

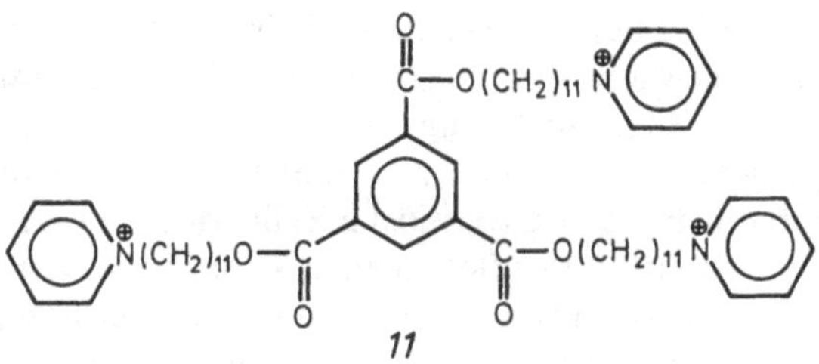

11

For example, high-field NMR data demonstrate that at equimolar quantities of (*11*) and phenol (3.3×10^{-3} M in acetonitrile), 85–90% of the phenol is bound. When phenol is in great excess over (*11*), several phenol molecules interact with each tentacle molecule. Binding of *p*-nitrophenolate to (*11*) in acetonitrile occurs with a huge association constant: 1.7×10^6 M^{-1}! Suckling also found that when phenol is enmeshed in the arms of (*11*), the phenol becomes resistant to the normally rapid chlorination by *t*-butyl hypochlorite. Protection of labile compounds, such as drugs, through encapsulation constitutes only one of several potential uses of multi-armed systems.

"Hexapus" (*12*), developed in our laboratory [10], also falls into the category of multi-armed compounds with ion-terminated chains. Six $[CH_2]_{10}CO_2^-$ chains project from a cyclotriveratrylene framework. Surprisingly, hexapus exhibits much less surface activity than single-chained fatty acid anions. Aqueous solutions of hexapus do not foam, and they possess surface tensions only slightly smaller than that of pure water. Apparently, the tendency of hexapus to absorb at the air-water interface is impaired by the difficulty of placing above the water phase *both* the hydrocarbon portion of the tails and the aromatic "cap".

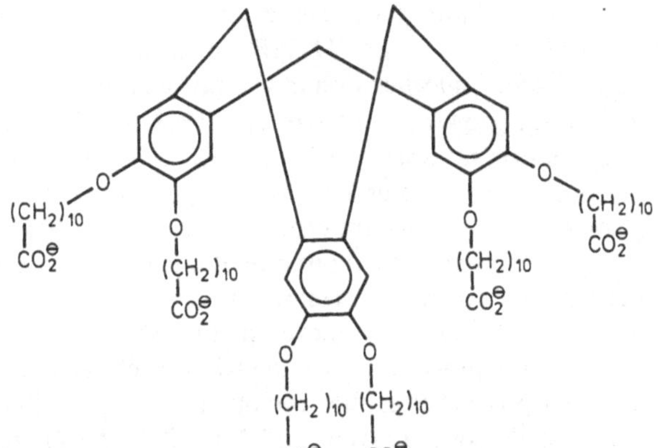

12

Hexapus in water solubilizes cholesterol, phenol blue ($K_{assoc} = 1.0 \times 10^4 M^{-1}$), naphthalene, and hydrophobic esters. Thus, hexapus seems non-selective in its binding characteristics (just like micelles). "Universal" binding has the advantage that almost any water-insoluble compound can be "collected" by the host molecule without regard to subtle structural variations. On the other hand, potential catalysts based on hexapus and other multi-armed systems would not be expected to manifest high specificity. Flexible chains do not lend themselves to a precise fit.

We were curious as to whether hexapus, where six chains are tied together covalently, would form aggregates in water as do single-chained fatty acid anions above a critical micelle concentration. Light scattering data show that hexapus does indeed assemble into small aggregates of about 9 for a total of 54 chains. Neither light scattering nor UV spectrophotometry reveals a critical micelle concentration for hexapus; if there is one, it must be extremely low (less than 1×10^{-5} M). At present we do not know whether guest molecules bind among the chains of a single hexapus molecule or among the chains of several hexapus molecules within an aggregate. Whichever the case, it is clear that hexapus has a distinct advantage over micelles: binding occurs at *all* concentrations, not just above a certain critical concentration.

We also investigated [11] three "trigapus" molecules (*13*)–(*15*). By themselves, the trigapus molecules are fairly mundane. Unlike ω-phenylalkanoic acids, they have no

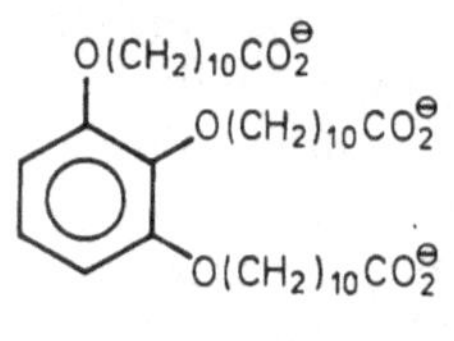

13

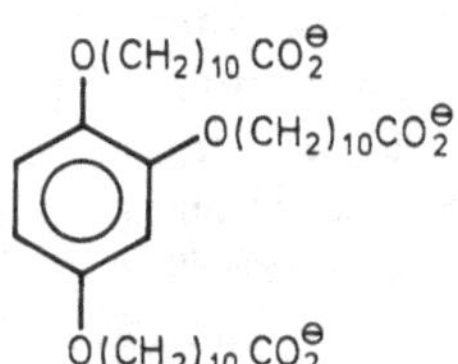

14

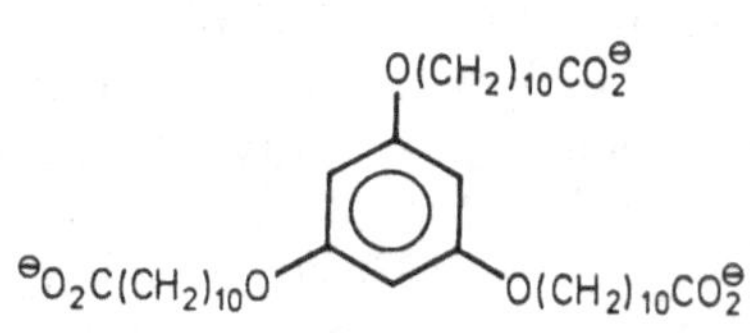

15

critical micelle concentration, form only very small aggregates barely detectable by light scattering, and do not bind small organics in water. Covalent attachment of the chains obviously has a dramatic effect on the colloidal behavior of amphiphilic molecules. The most interesting properties of the trigapus molecules relate to their interaction with cationic surfactants such as decyltrimethylammonium bromide (DTAB). Small amounts of trigapus lower DTAB's critical micelle concentration 10-fold. It is as if trigapus "seeds" micellization of the cationic surfactant. Low levels of trigapus also induce a huge growth in the size of DTAB micelles; thus, 1.3 mM trigapus elevates the aggregation number of DTAB micelles by at least an order of magnitude. Since these effects are not observed with trigapus and anionic surfactants, we presume that electrostatic attractions are critical to the phenomena.

Murakami and co-workers [12] have carried out one of the most thorough investigations of multi-armed compounds with ion-terminated chains. In 1979 they reported the substrate-binding behavior of an azaparacyclophane (*16*) in which the hydrophobic cavity was deepened by substitution of long ion-terminated chains on the macrocyclic skeleton. Salient properties of the cyclophane (*16*) include: (1) The compound has a critical micelle concentration of 3.2×10^{-4} M. (2) (*16*) binds cationic and neutral dyes but not anionic ones. Thus, Rhodamine 6G and Quinaldine Red form 1:1 complexes with (*16*) having association constants of about $5 \times 10^3 M^{-1}$.

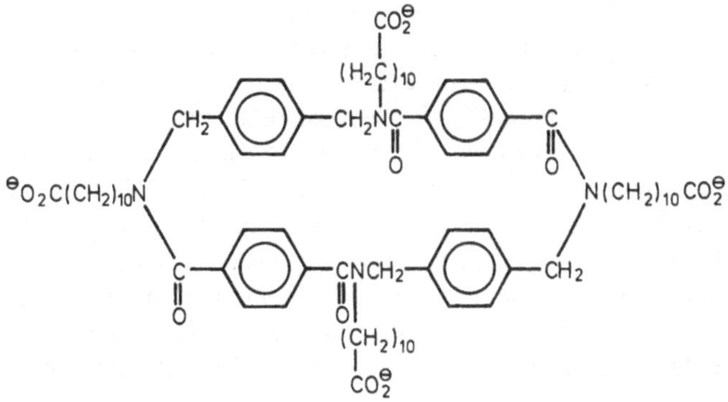

16

These bulky guests presumably reside in a cavity surrounded by the four aromatic rings of the cyclophane and the four alkyl branches. Since guests are incorporated into (*16*) below its critical micelle concentration, monomeric (*16*) provides an effective binding site apart from micellization processes. (3) Binding of *p*-nitrophenyl 3,5-dimethylcyclohexylacetate to (*16*) inhibits the basic hydrolysis of the former by a factor of 147 relative to the rate without (*16*). The rate data obey "reverse" Michaelis-Menten kinetics. (4) Hydrophobic spin probes, but not hydrophilic ones, bind to (*16*) and, as a consequence, have their rotational correlation times increased by as much as 3-fold. Hyperfine splitting constants do not, however, change on binding, suggesting that (*16*) associates only with the hydrophobic portion of the spin label while the nitroxide moiety remains outside in the water.

One can readily imagine multi-armed compounds bearing catalytically active groups; such materials could conceivably emulate enzymes by binding substrates prior to an intramolecular-type catalytic process. An example of this sort, provided by Murakami and co-workers [13], is shown under (*17*). The compound has two long alkyl chains terminated by cationic nitrogens; two other chains have, in addition, an imidazole ring with well-known catalytic properties toward ester hydrolysis. No

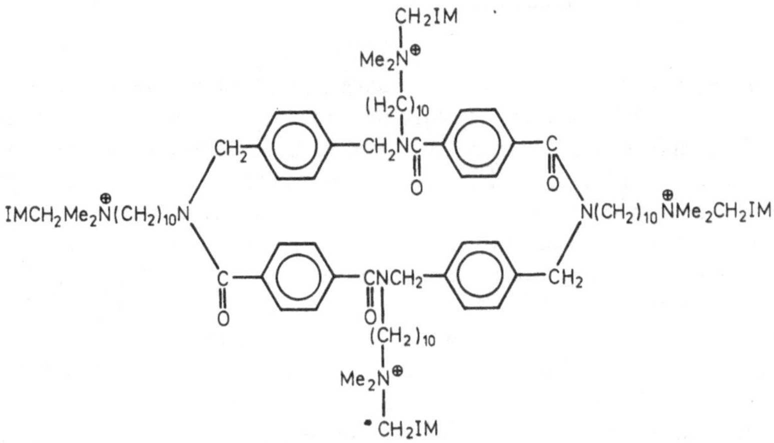

17

catalysis is observed with *p*-nitrophenyl esters of acids having eight carbons or less. On the other hand, fatty acid esters with chains of 10–16 carbons do indeed experience a catalyzed hydrolysis in the presence of (*17*). Since the C_{14} substrate reacts the fastest among the esters, the authors state, "There must be a certain threshold for the length of the alkyl chain of the substrate in order to attain the most favorable arrangement of both the ester bond of an incorporated substrate and the catalystic group (or groups) of an octopus cyclophane; this is achieved with the substrate, *p*-nitrophenyl tetradecanoate." Although the selectivity is really quite small (the C_{10}-C_{16} esters differ by less than a factor of 3), the concept as embodied in the quote is an important one. Selectivity among similar compounds, and regioselectivity among similar functional groups in a single compound, remain a primary goal of all chemists engaged in the design of cavities.

4 Neutral Multi-armed Materials

Tsukube [14] reported in 1984 a "multi-armed cyclam" (*18*) which has the ability to transport NH_4^+ cation through a chloroform layer (a so-called "liquid membrane"). Metal ions such as K^+ are not transported under similar conditions, a selectivity generally unobserved with the common crown ethers and cryptates. Since the transport rate decreases substantially when the furan rings terminating the pendant arms are

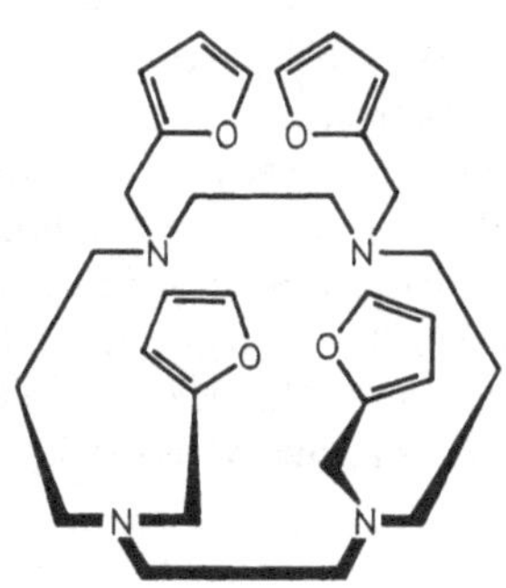

18

replaced by benzene rings, the furan oxygens must play an important role in the complexation of NH_4^+ cation. Tsukube feels that the ability of the multi-armed cyclam to differentiate NH_4^+ from K^+ is probably not related to ion-size (which is similar for the two ions). Instead, he suggests that charge distribution (being tetrahedral for NH_4^+ and spherical for K^+) is critical. CPK models indicate that the NH_4^+ cation can be wrapped tetrahedrally by its donating two hydrogen bonds to opposite ring nitrogens and two hydrogen bonds to furan oxygens. The furan-bearing crown (*19*), binds *both* NH_4^+ and K^+; substitution of benzene for furan has, in this case, only a minor effect on the transport properties of the system.

MacNicol and Wilson [15] synthesized a series of compounds (*20*) called "hexahosts". Such hexa-substituted benzenes can, on crystallization from suitable solvents, from a wide range of inclusion compounds. When Y = SPh, for example, a crystalline complex with CCl_4 was isolated having a host-guest stoichiometry of 1:2. The CCl_4 is

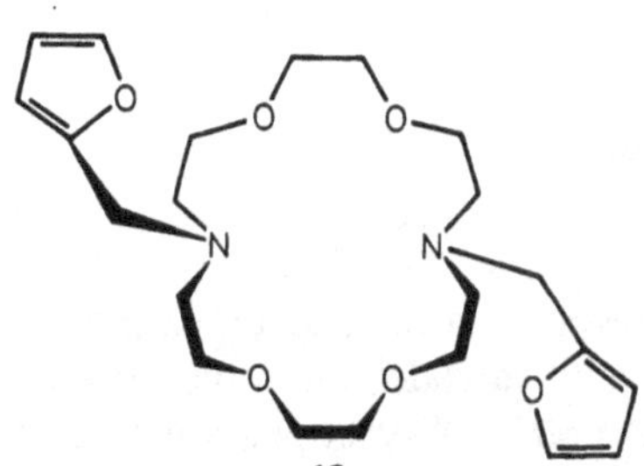

19

slowly lost upon standing. Whereas the Y = CH_2OPh and Y = CH_2SPh compounds both retain toluene, no such behavior was observed with Y = CH_2SePh. By far the largest number of inclusions compounds was obtained with the hexa-host having Y = $CH_2SC_6H_4$-*t*-Bu-*p*. Cyclooctane, phenyl iodide, bromoform, etc. all yield relatively stable crystalline complexes with this material.

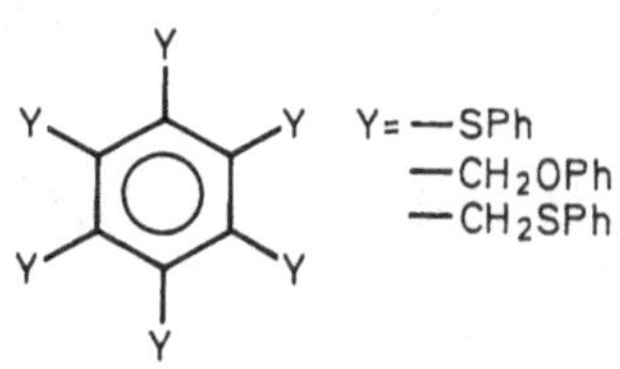

20

In more recent work, MacNicol and co-workers [16] described the first nitrogen-based hexa-host molecules (*21*). Although the derivative with R = $COCH_3$ (*21a*) shows no evidence of guest inclusion, the derivative with R = $COCF_3$ (*21b*) gives 1:2 host:guest adducts with nitromethane, tetramethylurea, N,N-dimethylformamide, etc. One of the most interesting observations pertains to the complex between the fluorine-containing host and the amide (*22*). Only the thermodynamically *less* stable *Z*-form of the amide incorporates into the crystals. Thus, the host displays complete configurational selectivity!

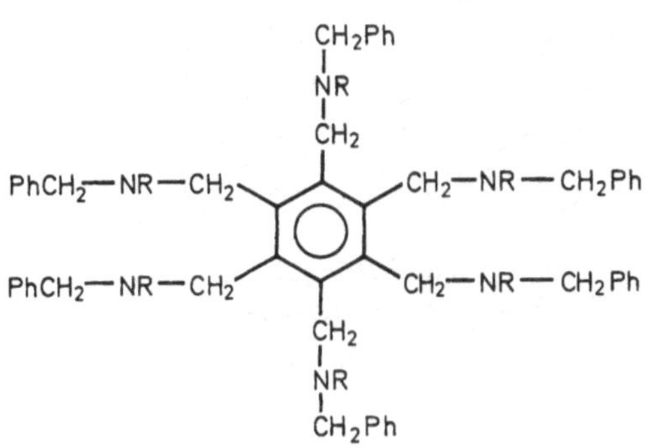

21 *a* R = $COCH_3$
b R = $COCF_3$

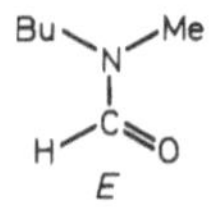

22

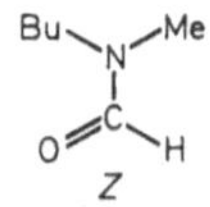

Newkome and co-workers [17] have recently developed an entirely new approach to multi-armed systems in their synthesis of "cascade" molecules. Rather than attaching chains to a central unit, they utilized a series of reactions which converted a single functional group into three functional groups. Each of the new functional groups can, in turn, be converted into three groups for a total of nine. Such a strategy could be continued to construct increasingly complex "tree-like" materials. A specific example, the synthesis of "[27]-arborol" (*22*), is given in Scheme 1. No doubt, intriguing new molecules are on the horizon.

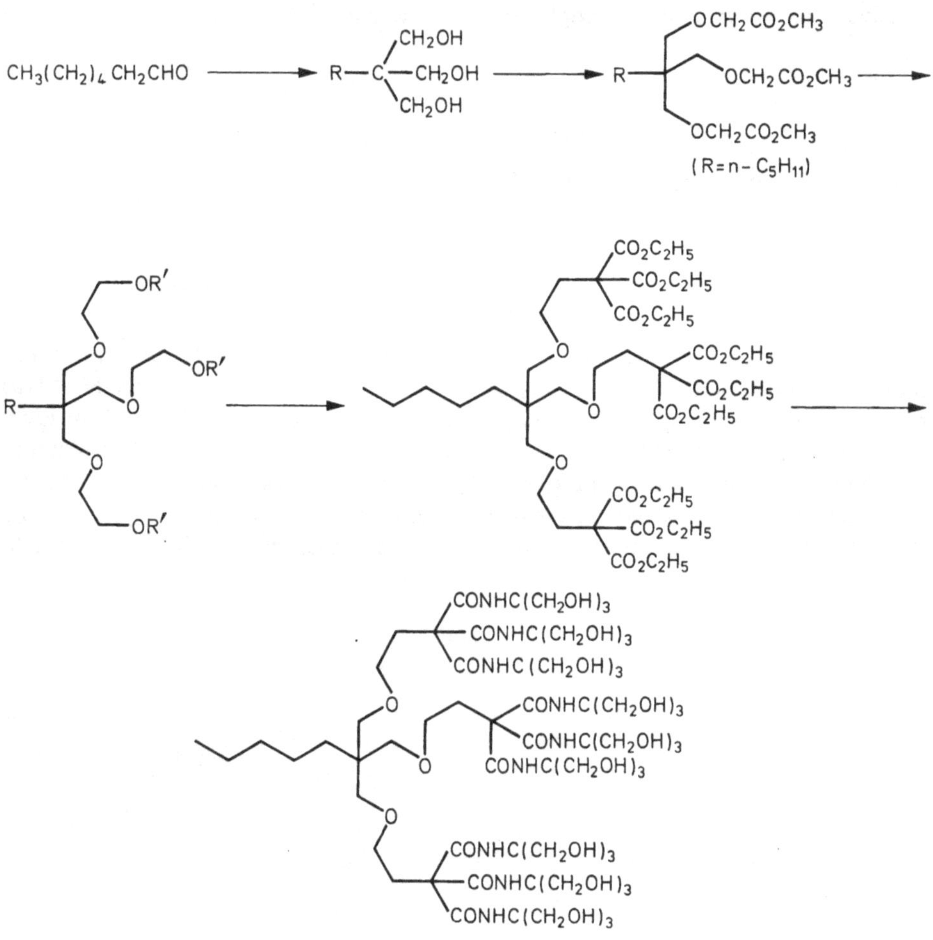

5 Polymer Systems

Thus far the discussion here has focused on compounds having multiple arms radiating from a small central unit such as a benzene, cyclotriveratrylene, or cyclophane ring. The number of such systems in the literature is, as yet, rather small. In contrast, there exists a vast body of data on polymeric chains bearing "arms" of various lengths. I have included a section on these polymers but obviously not with the intention of even superficially covering the subject. I merely wish to present four examples so that the interested organic and bio-organic chemist may, for comparison purposes, see how "arms" behave when they are covalently linked to a macromolecular backbone.

Almost three decades ago, Strauss and co-workers [18] carried out classic work on "polysoaps" (*23*) comprised of poly-4-vinylpyridine which had been quaternized on up to 38% of the nitrogens with *n*-dodecyl bromide; those nitrogens that escaped reaction with *n*-dodecyl bromide were then derivatized with ethyl bromide. Now addition of KBr to an aqueous solution of a polyelectrolyte normally decreases the viscosity. This is not true for the 38% polysoap (*23*) where the viscosity increases with

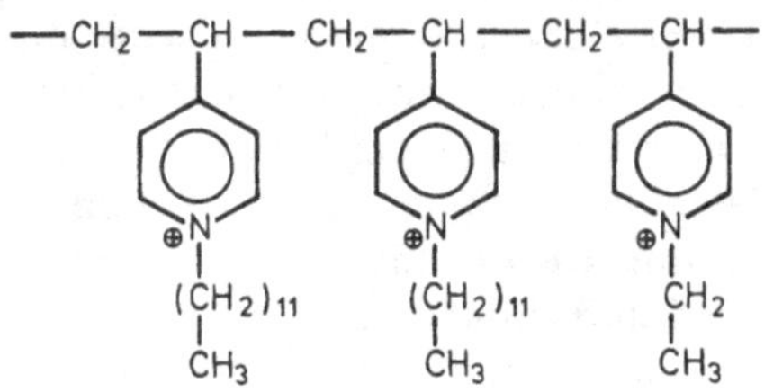

23

KBr until the solutions eventually gel. Viscosity data indicate that the polysoaps are far more compact than the random coils associated with ordinary polyelectrolytes in solution. In addition, the 38% polysoap (*23*) forms intermolecular aggregates in solution owing to the "sticky" hydrophobic spots on its surface. Disaggregation induced by dilution is a slow process (taking several hours). If one decreases the percentage of dodecyl groups on the polymer (relative to ethyl groups), then it is possible to revert back from polysoap behavior to that of a typical polyelectrolyte. This change occurs over a rather narrow composition range, suggesting the existence of a "critical dodecyl group content" analogous to the critical micelle concentration for simple soaps.

Cordes and co-workers [19] found that the alkaline hydrolysis of *p*-nitrophenyl hexanoate is subject to catalysis by polyvinylpyridine-based polysoaps. For example, k_{obs} is increased from 0.1 min^{-1} to 1.4 min^{-1} in the presence of 5×10^{-7} M 38% polysoap (*23*) (the same material used in the Strauss work). With 5×10^{-7} M polymer having a 15% dodecyl content, the rate is increased only 3 times above background. The simplest rationale for the kinetics invokes both hydrophobic and electrostatic forces. Thus, dodecyl chains on the polymer hydrophobically bind *p*-nitrophenyl hexanoate to the polymer surface. Since the polymer possesses a high density of cationic nitrogens, hydroxide ions also accumulate at the polymer surface where they catalyze the hydrolysis of bound ester. Addition of nitrate ion to the aqueous reaction mixture converts the polysoap-catalyzed process into a polysoap-inhibited one presumably because the nitrate displaces hydroxide counterions from the active sites.

Klotz and co-workers [20] attached pendant butanoyl, hexanoyl, and dodecanoyl groups onto polyethylenimine (*24*), a highly branched water-soluble polymer containing approximately 25% primary and tertiary nitrogens and 50% secondary nitrogens. Roughly 8–10% of the residues of the polymer were acylated. The acyl-

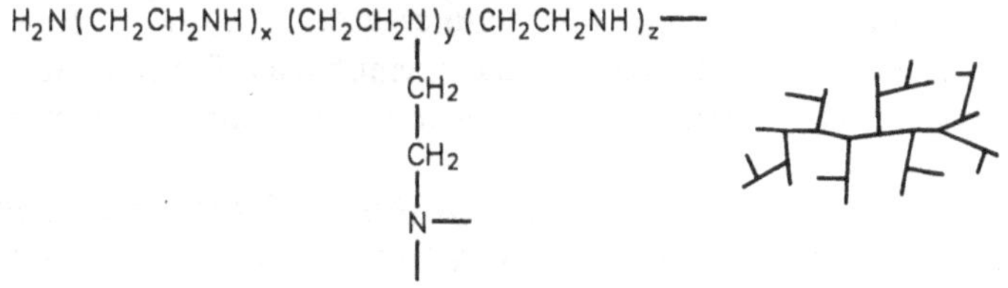

polyethylenimines are vastly more effective in binding a dye, methyl orange, than is serum albumin under comparable conditions. Thus, at a free methyl orange concentration of 10^{-5} M, the dodecanoyl, hexanoyl, and butanoyl derivatives bind 100, 10, and 1 dye molecules, respectively, compared to a value less than unity for bovine serum albumin. The improvement over albumin is impressive since the protein contains nearly 40% of nonpolar residues (although obviously not as long as a dodecyl chain). Urea (9.0 M) markedly reduces the binding affinity of the dodecanoyl-polyethylenimine. The dependence of binding on the dye concentration shows a strong cooperative interaction. In other words, each methyl orange anion creates, as it is bound, a new strong apolar site for additional binding.

Finally, mention should be made of the polysoap-catalyzed decarboxylation of 6-nitrobenzisoxazole-3-carboxylate anion (*Eq. 1*) studied by Kunitake and co-workers [21]. This reaction is known to proceed faster in apolar solvents than polar ones.

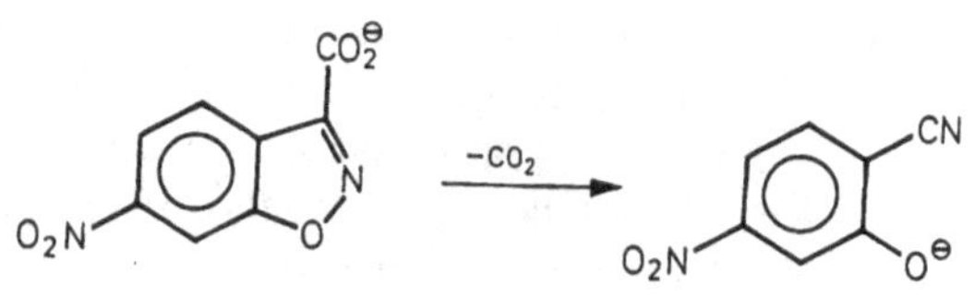

The polymers employed were (among others) partially dodecylated poly(2-ethyl-1-vinylimidazole) (*25*). It was found that polymer containing 29% dodecyl groups and 67% ethyl groups catalyzes the decarboxylation 350-fold, whereas a polymer with 9% dodecyl groups and 83% ethyl groups does not display significant catalysis. This kinetic behavior parallels the spectral shifts of bound methyl orange. When the dye

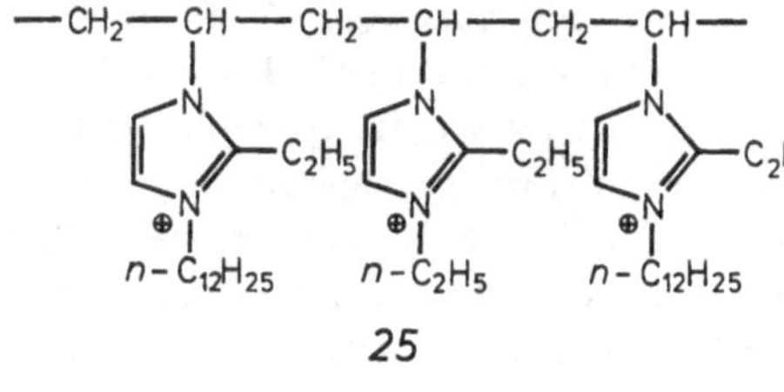

25

is bound to the 29% polymer, the λ_{max} shifts from 465 nm to 417 nm, attributable to a microenvironment less polar than water. In contrast, the catalytically inactive 9% polymer hardly perturbs the spectrum of methyl orange. Since the catalytic efficacy of the 29% polymer exceeds that of a conventional cationic micelle, tying chains together covalently can contribute positively to a rate process.

6 Concluding Remarks

Attaching multiple arms to a central unit has been shown to impart interesting chemical and physical properties not always predictable from the properties of the individual arms. Since the field of multi-armed compounds is relatively undeveloped, there is obviously much room for the imaginations of synthetically and physically inclined chemists to wander freely.

7 Acknowledgement

This work was supported by the National Institutes of Health and the Petroleum Research Fund.

8 References

1. Menger, F. M.: Acc. Chem. Res. *12*, 111 (1979)
2. Menger, F. M., Portnoy, C. E.: J. Am. Chem. Soc. *90*, 1875 (1968)
3. Bunton, C. A., Robinson, L., Stam, M. F.: Tetrahedron Lett. 121 (1971)
4. Vögtle, F., Weber, E.: Angew. Chem. Int. Ed. Engl. *13*, 814 (1974)
5. Vögtle, F., Müller, W. M., Buhleier, E., Wehner, W.: Chem. Ber. *112*, 899 (1979)
6. Hyatt, J. A.: J. Org. Chem. *43*, 1808 (1978)
7. Fornasier, R., Montanari, F.: Tetrahedron Lett. 1381 (1976)
8. Weber, E.: Angew. Chem. Int. Ed. Engl. *22*, 616 (1983)
9. Suckling, C. J.: J. Chem. Soc., Chem. Commun. 661 (1982)
10. Menger, F. M., Takeshita, M., Chow, J. F.: J. Am. Chem. Soc. *103*, 5938 (1981)
11. Menger, F. M., Angel de Greiff, A. J., Jaeger, D. A.: J. Chem. Soc., Chem. Commun. 543 (1984)
12. Murakami, Y., Nakano, A., Miyata, R., Matsuda, Y.: J. Chem. Soc., Perkin I, 1669 (1979)
13. Murakami, Y., Nakano, A., Akiyoshi, K., Fukuya, K.: ibid 2800 (1981)
14. Tsukube, H.: Chem. Lett. 1961 (1984)
15. MacNicol, D. D., Wilson, D. R.: J. Chem. Soc., Chem. Commun. 494 (1976)
16. Freer, A. A., Gall, J. H., MacNicol, D. D.: ibid. 674 (1982)
17. Newkome, G. R., Yao, Z., Baker, G. R., Gupta, V. K.: to be published in J. Org. Chem.
18. Strauss, U. P., Gershfeld, N. L., Crook, E. H.: J. Phys. Chem. *60*, 577 (1956)
19. Rodulfo, T., Hamilton, J. A., Cordes, E. H.: J. Org. Chem. *39*, 2281 (1974)
20. Klotz, I. M., Royer, G. P., Sloniewsky, A. R.: Biochem. *8*, 4752 (1969)
21. Kunitake, T., Shinkai, S., Hirotsu, S.: J. Org. Chem. *42*, 306 (1977)

Calculation of Interaction Energies in Host-Guest Systems

Martin Badertscher, Martin Welti, Pius Portmann and Ernö Pretsch

Department of Organic Chemistry, Swiss Federal Institute of Technology, CH-8092 Zürich, Switzerland

Table of Contents

1 Introduction

There is a strongly growing interest in designing and synthesizing host compounds for selected guests [1]. Although results of simple model calculations [2-3] have been applied for the development of ligand molecules for many years [4,5], merely a minor part of these works is based on theoretical considerations. In most cases the design is mainly intuitive and the sole tools used are molecular models. More elaborate theoretical calculations have only been applied since a few years as a design aid [6].

In the last decade much experience was collected independently on the host-guest chemistry in the field of the theoretical description of one particular kind of host-guest interaction: ions as guests and small ligand molecules as hosts [7-10]. Possibilities and limitations of such calculations are quite well known now. In the first part of this article the corresponding results will be reviewed. On the basis of these fundamental calculations different models were suggested for the description of host-guest interaction energies of larger systems. These models will be presented in Chapter 3. Since, in most cases, hosts (and often also guests) may adopt different conformations in different complexes as well as in the uncomplexed state, the calculation of the interaction energies is of only limited use if the conformational energy is not considered. Therefore several approaches of conformational energy calculations will be discussed briefly.

The main purpose of the present paper is to help judging the current possibilities and limitations of theoretical calculations on host-guest systems. In the last Chapter we will try to summarize the topic from this point of view.

An enormous work done in the field of polypeptides and proteins by Scheraga and coworkers includes calculations of both conformational energies and enzyme-substrate interactions for this special class of compounds [11-13]. This kind of calculations on host-guest systems is well documented elsewhere and is not considered within the scope of this article.

2 Ab initio Calculations of Ion-Ligand Interaction Energies

2.1 Introduction

Various methods for the calculation of ion-ligand interaction energies were discussed in several excellent reviews [7-9]. Therefore only a brief summary is given here. In some cases classical electrostatic models may suffice to predict correctly relative interaction energies and equilibrium geometries [7]. This success seems to be a consequence of error compensation [7]. Using such an electrostatic model, the interaction energies of a water molecule with alkali metal cations were calculated [14] within a few percent deviation relative to the experimental values measured in the gas phase [15]. But for realistic ionophores only a qualitative use of this model can be made because appropriate parameters are lacking [2]. Molecular electrostatic potentials are accessible from quantum chemical calculations and have been used to calculate interaction energies of ions with small ligand molecules [16] (see also Sect. 3.3).

In a series of papers various semiempirical quantum chemical procedures were examined for their usefulness to describe ion-ligand interactions (for a review see [7]).

The results were not very successful in several respects: wrong geometries, non-realistic interaction energies and extremly overemphasized charge transfer were obtained [7]. These techniques are thus unsuitable for a reliable investigation of ion-ligand interactions.

2.2 The Supermolecule Approach

Interaction energies (E_{Int}) on the basis of *ab initio* calculations are in general evaluated according to the so-called supermolecule approach:

$$E_{Int} = E_{Complex} - E_{Host} - E_{Guest} \tag{1}$$

E denotes on the right hand of the expression the total energy of the system specified. The interaction energy is by several orders of magnitude smaller than the individual total energies. Some consequences of this fact are discussed in the following sections.

A large number of complexes of alkali metal, alkaline earth metal and ammonium cations has been studied using this approach (see Table 1). Much less effort has been made in the field of anions as guests. Some of these results are collected in Table 2.

In case of complexes consisting of more than two constituents it has been shown that three-body terms are of significant magnitudes [120,121] i.e.:

$$E_{Int}(ABC) \neq E_{Int}(AB) + E_{Int}(BC) + E_{Int}(AC) \tag{2}$$

2.3 Basis Sets

The reliability of the *ab initio* calculations depends heavily on the choice of the basis sets. In order to reduce computational demands very often small basis sets have to be applied. According to experience, some rules can be given for the compatibility of the basis set with a given problem [122]. It is suggested that for molecular geometry optimizations and for the description of ion-ligand interactions minimal basis sets might be sufficient. In contrast, large basis sets are necessary for the computation of weak intermolecular interactions [136].

Many results shown in Table 1 clearly indicate that too small basis sets like STO-3G or a minimal GLO basis are inadequate to calculate reliable interaction energies (see also [31]). Their application is however justified if only relative stabilities and approximate geometries are to be evaluated. False geometries have been obtained for hydrogen bonded systems using the STO-3G basis set [123,124].

It is also known that geometry optimizations with too small basis sets may lead to non-realistic geometries especially if torsion angles or pyramidal structures are concerned [122]. In any case well-balanced basis sets are absolutely necessary, i.e. the quality of the basis set should be similar for all atoms. A carefully selected small GTO basis set may give reliable results for the ion-ligand interaction energies [31,125,126]. However, error compensations are at least partly responsible for this success. Therefore an improvement of the basis set may lead to less accurate results [91]. Non-balanced small basis sets lead to large basis set superposition errors [126-128]. This error is caused by the fact that in the calculations of the complex the wave functions

Table 1. Ab initio SCF calculations on the complexes of alkali, alkaline earth metal and ammonium cations with uncharged ligands.[a] The ligands are ordered according to their molecular mass.

Ion	Ligand	Basis set[b] atom	type	uncontracted	contracted	Ligand geometry/ * Complex geometry[c]	R [pm]	—E(Int) [kJ/mole]	Ref.
Li^+	NH_3	H	GTO	(4)	[2]	* STO-4G geometry,	190	169	17)
		N	GTO	(9/5/1)	[4/2/1]	* not varied	Li—N		
		Li	GTO	(10)	[4]				
Li^+	NH_3	H, N	GTO	STO-3G		assumed	not	229	18)
		Li	GTO	6G3G			given		
Li^+	NH_3	all	GTO	3-21G		* fully optimized	195	235	19)
							Li—N		
Li^+	NH_3	all	GTO	6-31G*		* 3-21G geometry	195	188	19)
							Li—N		
Li^+	NH_3	H	GTO	(6/1)	[3/1]				
		N	GTO	(11/7/1)	[5/4/1]	experimental	200	161	20)
		Li	GTO	(7/1)	[3/1]		Li—N		
Li^+	NH_3	H, N	GTO	4-31G		experimental	195	201	21)
		Li	GTO	(7/1)	[3/1]		Li—N		
Li^+	NH_3	H	GTO	(4)	[3]				
		N	GTO	(9/5)	[5/3]	experimental	200	189	21)
		Li	GTO	(7/1)	[3/1]		Li—N		
Li^+	NH_3	H, N	GTO	6-31G*		experimental	200	177	21)
		Li	GTO	(7/1)	[3/1]		Li—N		
Li^+	NH_3	H, N	GTO	6-31G**		experimental	200	174	21)
		Li	GTO	(7/1)	[3/1]		Li—N		
Li^+	NH_3	H	GTO	(4/1)	[3/1]				
		N	GTO	(9/5/1)	[5/3/1]	experimental	200	167	21)
		Li	GTO	(7/1)	[3/1]		Li—N		
Li^+	NH_3	H	GTO	(3)	[1]				
		N	GTO	(7/3)	[2/1]	assumed	196	182	22)
		Li	GTO	6G3G			Li—N		
Li^+	NH_3	H, N	GTO	4-31G		experimental	193	212	23)
		Li	GTO	5-21G			Li—N		

Li^+	NH_3	H	GTO	(5/1)	[3/1]				
		N	GTO	(11/7/2)	[5/5/2]	assumed	197	169	24)
		Li	GTO	(10/3)	[4/3]		Li—N		
Li^+	NH_3	all	GTO	4-31G		experimental	201	213	25)
							Li—N		
Li^+	NH_3	H	GTO	(4)	[2]				
		N	GTO	(7/3)	[4/2]	experimental	198	224	26)
		Li	GTO	(7/2)	[4/2]		Li—N		
Li^+	NH_3	H	GLO	(5)	[1]				
		N	GLO	(10/5)	[3/1]	assumed	not	209	27)
		Li	GLO	(10/2)	[3/1]		given		
Li^+	NH_3	H	GTO	(3)	[1]				
		N	GTO	(7/3)	[2/1]	not specified	192	167	28)
		Li	GTO	(7)	[3]		Li—N		
Li^+	NH_3	H, N	GTO	4-31G					
		Li	GTO	(8/2)?	[3/1]	not specified	197	211	28)
							Li—N		
Li^+	2 NH_3	H	GLO	(5)	[1]				
		N	GLO	(10/5)	[3/1]	assumed	not	395	27)
		Li	GLO	(10/2)	[3/1]		given		
Li^+	4 NH_3	H	GTO	(3)	[1]				
		N	GTO	(7/3)	[2/1]	not specified	205	515	28)
		Li	GTO	(7)	[3]		Li—N		
Li^+	4 NH_3	H	GTO	(4)	[1]	STO-3G geometry			
		N	GTO	(7/3)	[2/1]	* tetragonal,	200	653	29)
		Li	GTO	(7)	[2]	* part. optimized	Li—N		
Li^+	5 NH_3	H	GTO	(4)	[1]	STO-3G geometry			
		N	GTO	(7/3)	[2/1]	* pentagonal,	200	723	29)
		Li	GTO	(7)	[2]	* part. optimized	Li—N		
Li^+	6 NH_3	H	GTO	(4)	[1]	STO-3G geometry			
		N	GTO	(7/3)	[2/1]	* hexagonal,	200	741	29)
		Li	GTO	(7)	[2]	* part. optimized	Li—N		
Li^+	H_2O	H	GTO	(4)	[2]		184	157	
		O		(9/5/1)	[4/2/1]	* from Ref. [32]	Li—O		17)
		Li		(10)	[4]				

Table 1. (continued)

Ion	Ligand	Basis set[b] atom	type	uncontracted	contracted	Ligand geometry/ * Complex geometry[c]	R [pm]	—E(Int) [kJ/mole]	Ref.
Li^+	H_2O	all	GTO	STO-4G		STO-4G geometry	183 Li—O	168	30)
Li^+	H_2O	H, O	GTO	STO-3G		experimental	170 Li—O	335	31)
		Li	GTO	STO-3G + 1 p orbital					
Li^+	H_2O	H, O	GTO	STO-3G		experimental	175 Li—O	200	31)
		Li	GTO	6G3G					
Li^+	H_2O	H	GTO	(3)	[1]				
		O	GTO	(7/3)	[2/1]	experimental	183 Li—O	169	31)
		Li	GTO	6G3G					
Li^+	H_2O	H	GTO	(6/1)	[2/1]				
		O	GTO	(11/7/1)	[4/3/1]	fully optimized	189 Li—0	143	32)
		Li	GTO	(7/1)	[3/1]				
Li^+	H_2O	H	GTO	(6/2)	[4/2]				
		O	GTO	(13/8/2/1)	[8/5/2/1]	assumed	184 Li—O	147	32)
		Li	GTO	(8/2/1)	[5/2/1]				
Li^+	H_2O	H	GTO	(6/2)	[2/2]				
		O	GTO	(11/7/2)	[4/3/2]	not specified	189 Li—O	147	33)
		Li	GTO	(7/2/1)	[5/2/1]				
Li^+	H_2O	H	GTO	(6/1)	[3/1]				
		O	GTO	(11/7/1)	[5/4/1]	fully optimized	189 Li—O	151	34)
		Li	GTO	(11/2)	[5/2]				
Li^+	H_2O	H	GTO	(6/1)	[3/1]				
		O	GTO	(11/7/1)	[5/4/1]	experimental	183 Li—O	151	35)
		Li	GTO	(11/2)	[5/2]				
Li^+	H_2O	H	GTO	(4/1)	[2/1]				
		O	GTO	(9/5/1)	[4/2/1]	not specified	185 Li—O	155	36)
		Li	GTO	(9/3)	[4/2]				
Li^+	H_2O	all	GTO	3-21G		* fully optimized	177 Li—O	238	19)

Li^+	H_2O	all	GTO 6-31G*		* 3-21G geometry	177	163	19)
						Li—O		
Li^+	H_2O	H	GTO (6/1)	[3/1]				
		O	GTO (11/7/1)	[5/4/1]	experimental	180	141	20)
		Li	GTO (7/1)	[3/1]		Li—O		
Li^+	H_2O	H	GTO (3)	[1]				
		O	GTO (7/3)	[2/1]	assumed	183	170	22)
		Li	GTO 6G3G			Li—O		
Li^+	H_2O	H, O	GTO 4-31G		experimental	181	200	23)
		Li	GTO 5-21G			Li—O		
Li^+	H_2O	H	GTO (3)	[3]?				
		O	GTO (7/3)	[7/3]?	assumed	178	190	37)
		Li	GTO (7)	[7]?		Li—O		
Li^+	H_2O	H	GTO (3)	[3]?				
		O	GTO (7/3)	[7/3]?	assumed	181	137	37)
		Li	GTO (7/2)	[7/2]?		Li—O		
Li^+	H_2O	H	GTO (3/1)	[2/1]				
		O	GTO (7/3/2)	[4/2/2]?	assumed	184	163	37)
		Li	GTO (7/2)	[4/2]		Li—O		
Li^+	H_2O	all	GTO 4-31G		experimental	180	200	38)
						Li—O		
Li^+	H_2O	H	GTO (3)	[1]				
		O	GTO (9/10)	[2/1]	experimental	181	198	39)
		Li	GTO (9/2)	[2/1]		Li—O		
Li^+	H_2O	H	GTO (3)	[3]				
		O	GTO (9/10)	[4/2]	experimental	182	195	39)
		Li	GTO (9/2)	[4/1]		Li—O		
Li^+	H_2O	H	GTO (4)	[4]				
		O	GTO (9/10)	[6/4]	experimental	183	182	39)
		Li	GTO (9/2)	[7/1]		Li—O		
Li^+	H_2O	H	GTO (4)	[3]				
		O	GTO (9/5)	[5/3]	experimental	183	182	40)
		Li	GTO (7/1)	[3/1]		Li—O		
Li^+	H_2O	H	GTO (3)	[1]				
		O	GTO (7/3)	[2/1]	not specified	182	156	28)
		Li	GTO (7)	[3]		Li—O		

Table 1. (continued)

Ion	Ligand	Basis set[b] atom	type	uncontracted	contracted	Ligand geometry/ * Complex geometry[c]	R [pm]	−E(Int) [kJ/mole]	Ref.
Li^+	H_2O	H, O	GTO	4-31G					
		Li	GTO	(8/2)?	[3/1]	not specified	182 Li—O	195	[28]
Li^+	H_2O	H	GLO	(4/1)	[2/1]				
		O	GLO	(8/4/1)	[4/2/1]	* fully optimized	185 Li—O	145	[41]
		Li	GLO	(8/1)	[4/1]				
Li^+	H_2O	H	GLO	(1)	[1]				
		O	GLO	(2/1)	[2/1]	* fully optimized	183 Li—O	100	[41]
		Li	GLO	(2)	[1]				
Li^+	H_2O	H	GLO	(1)	[1]				
		O	GLO	(2/1)	[2/1]	experimental	183 Li—O	137	[41]
		Li	GLO	(2)	[1]				
Li^+	H_2O	all	GTO	6-31G**		* 6-31G* geometry	not given	166	[42]
Li^+	H_2O	H, Li	GTO	6-31G**		* 6-31G* geometry	not given	151	[42]
		O	GTO	6-31G** + sp orbital					
Li^+	2 H_2O	H	GTO	(6/1)	[3/1]				
		O	GTO	(11/7/1)	[5/4/1]	experimental	192 Li—O	283	[43]
		Li	GTO	(11/3)	[5/2]				
Li^+	2 H_2O	H	GTO	(3)	[2]				
		O	GTO	(7/3)	[4/2]	assumed	179 Li—O	370	[37]
		Li	GTO	(7)	[4]				
Li^+	2 H_2O	all	GTO	4-31G		* not optimized	180 Li—O	390	[44]
Li^+	4 H_2O	H	GTO	(3)	[1]				
		O	GTO	(7/3)	[2/1]	not specified	182 Li—O	494	[28]
		Li	GTO	(7)	[3]				
Li^+	4 H_2O	H	GTO	(4)	[1]	STO-3G geometry			
		O	GTO	(7/3)	[2/1]	* tetragonal,	185	608	[29]
		Li	GTO	(7)	[2]	* part. optimized	Li—O		

Li^+	5 H_2O	H	GTO (4)	[1]	STO-3G geometry			
		O	GTO (7/3)	[2/1]	* pentagonal,	190	671	29)
		Li	GTO (7)	[2]	* part. optimized	Li—O		
Li^+	6 H_2O	H	GTO (4)	[1]	STO-3G geometry			
		O	GTO (7/3)	[2/1]	* hexagonal,	200	680	29)
		Li	GTO (7)	[2]	* part. optimized	Li—O		
Li^+	HF	all	GTO 3-21G		* fully optimized	168 Li—F	183	19)
Li^+	HF	all	GTO 6-31G*		* 3-21G geometry	168 Li—F	117	19)
Li^+	HF	H, F	GTO 4-31G		experimental	175	144	23)
		Li	GTO 5-21G			Li—F		
Li^+	HCN	all	GTO 3-21G		* fully optimized	191 Li—N	194	19)
Li^+	HCN	all	GTO 6-31G*		* 3-21G geometry	191 Li—N	156	19)
Li^+	HCN	H	GTO (3/1)	[2/1]				
		C, N	GTO (7/3/1)	[4/2/1]	not specified	194	164	7)
		Li	GTO (6/1)	[3/1]		Li—N		
Li^+	HCN	H	GTO (4/1)	[3/1]				
		C, N	GTO (9/5/1)	[5/3/1]	not specified	196	155	7)
		Li	GTO (6/1)	[3/1]		Li—N		
Li^+	$CH_2=NH$	all	GTO 6-31G*		* 3-21G geometry	191 Li—N	189	45)
Li^+	HN=NH	all	GTO 6-31G*		* 3-21G geometry	196 Li—N	138	45)
Li^+	H_2CO	H	GTO (4)	[3]				
		C, O	GTO (9/5)	[5/3]	from Ref. 47)	177	181	46)
		Li	GTO (9)	[5]		Li—O		
Li^+	H_2CO	H	GTO (2)	[2]				
		C, O	GTO (5/3)	[3/2]	from Ref. 47)	180	188	46)
		Li	GTO (4)	[3]		Li—O		
Li^+	H_2CO	all	GTO 3-21G		* fully optimized	173 Li—O	222	19)
Li^+	H_2CO	all	GTO 6-31G*		* 3-21G geometry	173 Li—O	168	19)

Table 1. (continued)

Ion	Ligand	Basis set[b]				Ligand geometry/ * Complex geometry[c]	R [pm]	—E(Int) [kJ/mole]	Ref.
		atom	type	uncontracted	contracted				
Li^+	H_2CO	H, C, O	GTO	4-31G		experimental	176	198	23)
		Li	GTO	5-21G			Li—O		
Li^+	H_2CO	all	GTO	STO-3G		* fully optimized	not given	308	48)
Li^+	H_2CO	H	GLO	(5)	[2]				
		C, O	GLO	(10/5)	[4/2]	* part. optimized	180	174	49)
		Li	GLO	(10)	[4]		Li—O		
Li^+	H_2CO	H	GLO	(5)	[2]				
		C, O	GLO	(10/5)	[4/2]	* part. optimized	180	171	49)
		Li	GLO	(10)	[4]		Li—O		
		counterpoise corrected							
Li^+	H_2O	H	GTO	(4)	[3]				
		C, O	GTO	(9/5)	[5/3]	assumed	178	184	50)
		Li	GTO	(7/1)	[3/1]		Li—O		
Li^+	H_2CO	all	GTO	STO-3G		* fully optimized	not given	404	51)
Li^+	H_2CO	H	GTO	(4)	[3]				
		C, O	GTO	(9/5/1)	[5/3/1]	assumed	182	161	40)
		Li	GTO	(7/1)	[3/1]		Li—O		
Li^+	H_2CO	all	GTO	6-31G		* fully optimized	177	185	52) 53)
							Li—O		
Li^+	$4\ H_2CO$	H	GTO	(4)	[1]	STO-3G geometry			
		C, O	GTO	(7/3)	[2/1]	* tetragonal,	185	489	29)
		Li	GTO	(7)	[2]	* part. optimized	Li—O		
Li^+	$5\ H_2CO$	H	GTO	(4)	[1]	STO-3G geometry			
		C, O	GTO	(7/3)	[2/1]	* pentagonal,	200	524	29)
		Li	GTO	(7)	[2]	* part. optimized	Li—O		
Li^+	$6\ H_2CO$	H	GTO	(4)	[1]	STO-3G geometry			
		C, O	GTO	(7/3)	[2/1]	* hexagonal,	210	520	29)
		Li	GTO	(7)	[2]	* part. optimized	Li—O		

Li^+	$CH_3—NH_2$	all	GTO	3-21G		*	fully optimized	195 Li—N	235	[19]
Li^+	$CH_3—NH_2$	all	GTO	6-31G*		*	3-21G geometry	195 Li—N	187	[19]
Li^+	$CH_3—NH_2$	H, C, N Li	GTO GTO	STO-3G 6G3G			assumed	not given	220	[18]
Li^+	$CH_3—NH_2$	H, C, N Li	GTO GTO	4-31G (7/1)	 [3/1]		experimental	195 Li—N	200	[21]
Li^+	$CH_3—NH_2$	H, C, N Li	GTO GTO	6-31G* (7/1)	 [3/1]		experimental	200 Li—N	175	[21]
Li^+	$CH_3—NH_2$	H, C, N Li	GTO GTO	6-31G** (7/1)	 [3/1]		experimental	200 Li—N	173	[21]
Li^+	$CH_3—NH_2$	H C, N Li	GTO GTO GTO	(4/1) (9/5/1) (7/1)	[3/1] [5/3/1] [3/1]		experimental	200 Li—N	167	[21]
Li^+	$CH_3—NH_2$	all	GTO	4-31G			experimental	not given	204	[25]
Li^+	$HN=O$	all	GTO	6-31G*		*	3-21G geometry	182 Li—O	113	[45]
Li^+	$NH_2—NH_2$	all	GTO	6-31G*		*	3-21G geometry	193 Li—N	199	[45]
Li^+	$CH_3—OH$	all	GTO	3-21G		*	fully optimized	176 Li—O	243	[19]
Li^+	$CH_3—OH$	all	GTO	6-31G*		*	3-21G geometry	176 Li—O	172	[19]
Li^+	$CH_3—OH$	H C, O Li	GTO GTO GTO	(3) (7/3) 6G3G	[1] [2/1] 		assumed	not given	169	[22]
Li^+	$CH_3—OH$	H C, O Li	GTO GTO GTO	(3) (7/3) (7)	[1] [2/1] [3]		not specified	180 Li—O	154	[28]
Li^+	$CH_3—OH$	H, C, O Li	GTO GTO	4-31G (8/2)?	 [3/1]		not specified	181 Li—O	203	[28]
Li^+	$NH_2—OH$	all	GTO	6-31G*		*	3-21G geometry	178 Li—O	177	[45]

Table 1. (continued)

Ion	Ligand	Basis set[b]				Ligand geometry/ * Complex geometry[c]	R [pm]	—E(Int) [kJ/mole]	Ref.
		atom	type	uncontracted	contracted				
Li^+	HO—OH	all	GTO	6-31G*		* 3-21G geometry	189 Li—O	156	45)
Li^+	CH_3—F	all	GTO	6-31G*		* 3-21G geometry	166 Li—F	146	45)
Li^+	PH_3	all	GTO	3-21G		* fully optimized	258	119	19)
Li^+	PH_3	all	GTO	6-31G*		* 3-21G geometry	258 Li—P	109	19)
Li^+	H_2S	H, S Li	GTO GTO	STO-3G (7/1)	 [3/1]	experimental	220 Li—S	184	7)
Li^+	H_2S	H, S Li	GTO GTO	4-31G (7/1)	 [3/1]	experimental	240 Li—S	92	7)
Li^+	H_2S	all	GTO	3-21G		* fully optimized	244 Li—S	98	19)
Li^+	H_2S	all	GOT	6-31G*		* 3-21G geometry	244 Li—S	75	19)
Li^+	H_2S	H, S Li	GTO GTO	4-31G 5-21G		experimental	249 Li—S	109	55)
Li^+	NH_2—F	all	GTO	6-31G*		* 3-21G geometry	182 Li—F	152	45)
Li^+	HO—F	all	GTO	6-31G*		* 3-21G geometry	178 Li—F	105	45)
Li^+	HCl	all	GTO	3-21G		* fully optimized	245 Li—Cl	68	19)
Li^+	HCl	all	GTO	6-31G*		* 3-21G geometry	245 Li—Cl	57	19)
Li^+	HCl	H, Cl Li	GTO GTO	4-31G 5-21G		experimental	241 Li—Cl	74	23)
Li^+	CH_3—CN	all	GTO	3-21G		* fully optimized	188 Li—N	226	19)

Li^+	CH_3—CN	all	GTO	6-31G*		* 3-21G geometry	188 Li—N	188	19)
Li^+	CH_3—CN	H C, N Li	GTO GTO GTO	(3) (7/3) 6G3G	[1] [2/1]	assumed	not given	169	22)
Li^+	CH_2–NH–CH_2 (ring)	all	GTO	6-21G		* fully optimized	189 Li—N	253	56)
Li^+	CH_2–NH–CH_2 (ring)	all	GTO	6-31G*		* fully optimized	196 Li—N	198	56)
Li^+	CH_3—CHO	all	GTO	3-21G		* fully optimized	171 Li—O	245	19)
Li^+	CH_3—CHO	all	GTO	6-31G*		* 3-21G geometry	171 Li—O	190	19)
Li^+	CH_2–O–CH_2 (ring)	all	GTO	6-21G		* fully optimized	172 Li—O	252	56)
Li^+	CH_2–O–CH_2 (ring)	all	GTO	6-31G*		* fully optimized	180 Li—O	182	56)
Li^+	CH_3—NH—CH_3	H, C, N Li	GTO GTO	STO-3G 6G3G		assumed	not given	213	18)
Li^+	CH_3—NH—CH_3	all	GTO	3-21G		* fully optimized	193 Li—N	231	19)
Li^+	CH_3—NH—CH_3	all	GTO	6-31G*		* 3-21G geometry	193 Li—N	182	19)
Li^+	CH_3—CH_2—NH_2	H C, N Li	GTO GTO GTO	(3) (7/3) 6G3G	[1] [2/1]	assumed	not given	177	22)
Li^+	HCO—NH_2	H C, N, O Li	GTO GTO GTO	(3) (7/3) 6G3G	[1] [2/1]	assumed	not given	212	22)
Li^+	HCO—NH_2	H C, N, O Li	GLO GLO GLO	(3) (4/2) (4/1)	[3] [4/2] [4/1]	experimental	172 Li—O	234	57)
Li^+	HCO—NH_2	H C, N, O Li	GLO GLO GLO	(1) (2/1) (2)	[1] [2/1] [1]	assumed	175 Li—O	171	58)

Table 1. (continued)

Ion	Ligand	Basis set[b] atom	type	uncontracted	contracted	Ligand geometry/ * Complex geometry[c]	R [pm]	−E(Int) [kJ/mole]	Ref.
Li^+	$HCO-NH_2$	H	GTO	(3)	[3]?				
		C, N, O	GTO	(6/3)	[4/2]	assumed	172 Li—O	219	58)
		Li	GTO	(6/1)	[4/1]				
Li^+	$HCO-NH_2$	H	GLO	(1)	[1]				
		C, N, O	GLO	(2/1)	[2/1]	experimental	177 Li—O	162	59)
		Li	GLO	(2)	[1]				
Li^+	$HCO-NH_2$	H	GLO	(4/1)	[2/1]				
		C, N, O	GLO	(8/4/1)	[4/2/1]	experimental	not given	218	59)
		Li	GLO	(8/1)	[4/1]				
Li^+	$2\,HCO-NH_2$	H	GLO	(3)	[3]				
		C, N, O	GLO	(4/2)	[4/2]	experimental	183 Li—O 219 Li—N	431	60)
		Li	GLO	(4/1)	[4/1]				
Li^+	CH_3-CH_2-OH	H	GTO	(3)	[1]				
		C, O	GTO	(7/3)	[2/1]	assumed	not given	175	22)
		Li	GTO	6G3G					
Li^+	CH_3-O-CH_3	all	GTO	3-21G		* fully optimized	175 Li—O	242	19)
Li^+	CH_3-O-CH_3	all	GTO	6-31G*		* 3-21G geometry	175 Li—O	172	19)
Li^+	CH_3-O-CH_3	all	GTO	STO-3G		experimental	170 Li—O	326	61)
Li^+	CH_3-O-CH_3	H	GTO	(3)	[1]				
		C, O	GTO	(7/3)	[2/1]	experimental	170 Li—O	131	62)
		Li	GTO	(5)	[1]				
Li^+	HCOOH	H	GLO	(1)	[1]				
		C, O	GLO	(2/1)	[2/1]	assumed	175 Li—O	171	63)
		Li	GLO	(2)	[1]				
Li^+	CH_3-PH_2	all	GTO	3-21G		* fully optimized	256 Li—P	136	19)

Li^+	CH_3—PH_2	all	GTO	6-31G*		* 3-21G geometry	256 Li—P	131	19)
Li^+	CH_3—SH	all	GTO	3-21G		* fully optimized	240 Li—S	121	19)
Li^+	CH_3—SH	all	GTO	6-31G*		* 3-21G geometry	240 Li—S	100	19)
Li^+	CH_3—Cl	all	GTO	3-21G		* fully optimized	233 Li—Cl	111	19)
Li^+	CH_3—Cl	all	GTO	6-31G*		* 3-21G geometry	233 Li—Cl	87	19)
Li^+	$(CH_3)_2CH$—CN	H C, N Li	GTO GTO GTO	(3) (7/3) 6G3G	[1] [2/1]	assumed	not given	158	22)
Li^+	CH_3—CO—CH_3	H C, O Li	GTO GTO GTO	(3) (7/3) 6G3G	[1] [2/1]	assumed	176 Li—O	181	22)
Li^+	CH_3—CO—CH_3	all	GTO	6-31G		* fully optimized	not given	222	52)
Li^+	HC=O—HC=O	H C, O Li	GLO GLO GLO	(4) (8/4) (8)	[3] [4/2] [4]	experimental	not given	197	64)
Li^+	HC=O—HC=O	H C, O Li	GLO GLO GLO	(1) (2/1) (2)	[1] [2/1] [1]	experimental	nct given	113	65)
Li^+	2 HC=O—HC=O	H C, O Li	GTO GTO GTO	(4) (7/3) (7/3)	[2] [4/2] [4/1]	assumed	202 Li—O	425	66)
Li^+	$N(CH_3)_3$	H, C, N Li	GTO GTO	STO-3G 6G3G		assumed	not given	201	18)
Li^+	$(CH_3)_2CH$—NH_2	H C, N Li	GTO GTO GTO	(3) (7/3) 6G3G	[1] [2/1]	assumed	not given	181	22)
Li^+	CH_3—CO—NH_2	H C, N, O Li	GLO GLO GLO	(1) (2/1) (2)	[1] [2/1] [1]	experimental	175 Li—O	178	67)

Table 1. (continued)

Ion	Ligand	Basis set[b] atom	type	uncontracted	contracted	Ligand geometry/ * Complex geometry[c]	R [pm]	—E(Int) [kJ/mole]	Ref.
Li^+	$HCO-NH-CH_3$	H	GTO	(3)	[1]				
		C, N, O	GTO	(7/3)	[2/1]	assumed	175	213	[22]
		Li	GTO	6G3G			Li—O		
Li^+	$HCO-NH-CH_3$	H	GLO	(1)	[1]				
		C, N, O	GLO	(2/1)	[2/1]	assumed	177	155	[68]
		Li	GLO	(2)	[1]		Li—O		
Li^+	$NH_2-CO-NH_2$	all	GTO	6-31G		* fully optimized	not given	271	[52]
Li^+	$HO-CO-OH$	all	GTO	6-31G		* fully optimized	not given	195	[52]
Li^+	$CH_3-PH-CH_3$	all	GTO	3-21G		* fully optimized	255 Li—P	148	[19]
Li^+	$CH_3-PH-CH_3$	all	GTO	6-31G*		* 3-21G geometry	255 Li—P	148	[19]
Li^+	CH_3-S-CH_3	all	GTO	3-21G		* fully optimized	237 Li—S	136	[19]
Li^+	CH_3-S-CH_3	all	GTO	6-31G*		* 3-21G geometry	237 Li—S	119	[19]
Li^+	CH_3-S-CH_3	H	GTO	(3)	[1]				
		C	GTO	(7/3)	[2/1]	experimental	243	70	[62]
		S	GTO	(9/6)	[3/2]		Li—S		
		Li	GTO	(5)	[1]				
Li^+	$F-CO-F$	all	GTO	6-31G		* fully optimized	not given	105	[52]
Li^+	$CHO-CH_2-CHO$	H	GLO	(1)	[1]				
		C, O	GLO	(2/1)	[2/1]	assumed	not	200	[65]
		Li	GLO	(2)	[1]		given		
Li^+	$(CH_3)_3C-NH_2$	H	GTO	(3)	[1]				
		C, N	GTO	(7/3)	[2/1]	assumed	not	182	[22]
		Li	GTO	6G3G			given		

Li^+	$CH_3-CH_2-CO-NH_2$	H C, N, O Li	GLO (1) GLO (2/1) GLO (2)	[1] [2/1] [1]	experimental	175 Li—O	187	67)
Li^+	$HCO-NH-CH_2-CH_3$	H C, N, O Li	GTO (3) GTO (7/3) GTO 6G3G	[1] [2/1]	assumed	not given	221	22)
Li^+	$HCO-N(CH_3)_2$	H C, N, O Li	GTO (3) GTO (7/3) GTO 6G3G	[1] [2/1]	assumed	not given	223	22)
Li^+	$HCO-N(CH_3)_2$	H C, N, O Li	GLO (1) GLO (2/1) GLO (2)	[1] [2/1] [1]	experimental	not given	169	69)
Li^+	$CH_3-CO-NH-CH_3$	H C, N, O Li	GLO (1) GLO (2/1) GLO (2)	[1] [2/1] [1]	experimental * not varied	175 Li—O	180	67)
Li^+	$CH_3-CH_2-O-CH_2-CH_3$	H C, O Li	GTO (3) GTO (7/3) GTO 6G3G	[1] [2/1]	assumed	not given	181	22)
Li^+	$CH_3-CH_2-O-CH_2-CH_3$	H C, O Li	GTO (3) GTO (7/3) GTO (5)	[1] [2/1] [1]	experimental	172 Li—O	135	62)
Li^+	$CH_3-COO-CH_3$	H C, O Li	GTO (3) GTO (7/3) GTO 6G3G	[1] [2/1]	assumed	not given	172	22)
Li^+	C_5H_5N (pyridine)	H C, N Li	GTO (3) GTO (7/3) GTO 6G3G	[1] [2/1]	assumed	192 Li—N	179	22)
Li^+	$CHO-CH_2-CH_2-CHO$	H C, O Li	GLO (1) GLO (2/1) GLO (2)	[1] [2/1] [1]	assumed	not given	236	65)
Li^+	$CH_3-CH_2-CH_2-CO-NH_2$	H C, N, O Li	GLO (1) GLO (2/1) GLO (2)	[1] [2/1] [1]	experimental	175 Li—O	191	67)
Li^+	$CH_3-CO-N(CH_3)_2$	H C, N, O Li	GTO (3) GTO (7/3) GTO (5)	[1] [2/1] [1]	assumed	169 Li—O	189	62)

Table 1. (continued)

Ion	Ligand	Basis set[b] atom	type	uncontracted	contracted	Ligand geometry/ * Complex geometry[c]	R [pm]	−E(Int) [kJ/mole]	Ref.
Li^+	$CH_3-CO-NH-CH_2-CH_3$	H	GTO	(3)	[1]				
		C, N, O	GTO	(7/3)	[2/1]	assumed	not given	231	22)
		Li	GTO	6G3G					
Li^+	$CH_3-CO-NH-CH_2-CH_3$	H	GLO	(1)	[1]	experimental			
		C, N, O	GLO	(2/1)	[2/1]	* not varied	175 Li—O	185	67)
		Li	GLO	(2)	[1]				
Li^+	$CH_3-CO-N(CH_3)_2$	H	GTO	(3)	[1]				
		C, N, O	GTO	(7/3)	[2/1]	assumed	not given	232	22)
		Li	GTO	6G3G					
Li^+	$CH_3-COO-CH_2-CH_3$	H	GTO	(3)	[1]				
		C, O	GTO	(7/3)	[2/1]	assumed	not given	178	22)
		Li	GTO	6G3G					
Li^+	$CH_3-O-CH_2-CH_2-O-CH_3$	H	GTO	(3)	[1]	assumed			
		C, O	GTO	(7/3)	[2/1]	synperiplanar	178 Li—O	244	70)
		Li	GTO	(5)	[1]	E(conf) = 30 kJ/mole			
Li^+	2 HC=S–HC=S (HC=S above HC=S, joined by a vertical bond)	H	GTO	(4)	[2]				
		C	GTO	(7/3)	[4/2]	assumed	213 Li—O	87	66)
		S	GTO	(10/6)	[5/4]				
		Li	GTO	(7/3)	[4/1]				
Li^+	$CH_3-(CH_2)_3-CO-NH_2$	H	GLO	(1)	[1]	experimental			
		C, N, O	GLO	(2/1)	[2/1]	* not varied	175 Li—O	193	67)
		Li	GLO	(2)	[1]				
Li^+	$HCO-N(CH_2-CH_3)_2$	H	GTO	(3)	[1]				
		C, N, O	GTO	(7/3)	[2/1]	assumed	not given	232	22)
		Li	GTO	6G3G					
Li^+	$NH_2-CO-NH-CO-NH_2$	H	GLO	(1)	[1]				
		C, N, O	GLO	(2/1)	[2/1]	experimental	180 Li—O	269	71)
		Li	GLO	(2)	[1]				

Li^+	$2\ NH_2{-}CO{-}NH{-}CO{-}NH_2$	H C, N, O Li	GLO (1) GLO (2/1) GLO (2)	[1] [2/1] [1]	experimental	not given	302	71)
Li^+	$CO{-}N{=}C(NH_2){-}CH{=}CH{-}NH$ (ring) cytosine	H C, N, O Li	GLO (1) GLO (2/1) GLO (2)	[1] [2/1] [1]	experimental	195 Li—O	270	72)
Li^+	$CO{-}N{=}C(NH_2){-}CH{=}CH{-}NH$ (ring) cytosine	H C, N, O Li counterpoise corrected	GLO (1) GLO (2/1) GLO (2)	[1] [2/1] [1]	experimental	195 Li—O	242	73)
Li^+	$CH_3{-}CO{-}N(CH_2{-}CH_3)_2$	H C, N, O Li	GTO (3) GTO (7/3) GTO 6G3G	[1] [2/1]	assumed	not given	238	22)
Li^+	$CH_3{-}CO{-}N(CH_2{-}CH_3)_2$	H C, N, O Li	GTO (3) GTO (7/3) GTO (5)	[1] [2/1] [1]	assumed	170 Li—O	193	62)
Li^+	$CO{-}NH{-}CH{=}C(CH_3){-}CO{-}NH$ (ring) thymine	H C, N, O Li	GLO (1) GLO (2/1) GLO (2)	[1] [2/1] [1]	experimental	176 Li—O	188	72)
Li^+	$CO{-}NH{-}CH{=}C(CH_3){-}CO{-}NH$ (ring) thymine	H C, N, O Li counterpoise corrected	GLO (1) GLO (2/1) GLO (2)	[1] [2/1] [1]	experimental	176 Li—O	168	73)
Li^+	$N{-}C(NH_2){=}C{-}N{=}CH$ $CH{-}{-}N{=}{=}C{-}{-}{-}NH$ adenine	H C, N Li	GLO (1) GLO (2/1) GLO (2)	[1] [2/1] [1]	experimental	185 Li—O	212	72)
Li^+	$N{-}C(NH_2){=}C{-}N{=}CH$ $CH{-}{-}N{=}{=}C{-}{-}{-}NH$ adenine	H C, N Li counterpoise corrected	GLO (1) GLO (2/1) GLO (2)	[1] [2/1] [1]	experimental	185 Li—O	194	73)
Li^+	$NH{-}CO{-}C{-}N{=}CH$ $NH_2{-}C{=}{=}N{-}{-}C{-}{-}{-}NH$ guanine	H C, N, O Li	GLO (1) GLO (2/1) GLO (2)	[1] [2/1] [1]	experimental	192 Li—O	293	72)

Table 1. (continued)

Ion	Ligand	Basis set[b] atom	type	uncontracted	contracted	Ligand geometry/ * Complex geometry[c]	R [pm]	−E(Int) [kJ/mole]	Ref.
Li^+	NH–CO–C–N=CH NH_2–C==N––C–––NH guanine	H C, N, O Li counterpoise corrected	GLO GLO GLO	(1) (2/1) (2)	[1] [2/1] [1]	experimental	192 Li—O	262	73)
Li^+	CH_2–NH–CH_2–CH_2–NH–CH_2 CH_2–NH–CH_2–CH_2–NH–CH_2	H C, N Li	GLO GLO GLO	(1) (2/1) (2)	[1] [2/1] [1]	assumed * part. optimized	188 Li—N	517	74)
Li^+	CH_3–NH–CH_2–CH_2–NH–CH_2 CH_3–NH–CH_2–CH_2–NH–CH_2	H C, N Li	GLO GLO GLO	(1) (2/1) (2)	[1] [2/1] [1]	assumed * not varied	188 Li—N	376	74)
Li^+	CH_2–O–CH_2–CH_2–O–CH_2 CH_2–O–CH_2–CH_2–O–CH_2	all	GTO	STO-3G		assumed * not optimized	180 Li—O	923	61)
Li^+	CH_2–O–CH_2–CH_2–O–CH_2 CH_2–O–CH_2–CH_2–O–CH_2	H C, O Li	GLO GLO GLO	(1) (2/1) (2)	[1] [2/1] [1]	assumed * part. optimized	204 Li—O	241	74)
Li^+	CH_3–O–CH_2–CH_2–O–CH_2 CH_3–O–CH_2–CH_2–O–CH_2	H C, O Li	GLO GLO GLO	(1) (2/1) (2)	[1] [2/1] [1]	assumed * not varied	204 Li—O	68	74)
Li^+	CH_2–(O–CH_2–CH_2)$_2$–O–CH_2 CH_2–(O–CH_2–CH_2)$_2$–O–CH_2	H C, O Li	GTO GTO GTO	(3) (7/3) (5)	[1] [2/1] [1]	geometry of complexed ligand * experimental	207 Li—O	317	75)
Na^+	NH_3	all	GTO	3-21G		* fully optimized	230 Na—N	172	19)
Na^+	NH_3	all	GTO	6-31G*		* 3-21G geometry	230 Na—N	136	19)
Na^+	NH_3	H N Na	GTO, GTO GTO	(6/1) (11/7/1) (12/6)	[3/1] [5/4/1] [6/4]	experimental	240 Na—N	116	20)

Na^+	4 NH_3	H	GTO	(4)	[1]	STO-3G geometry			
		N	GTO	(7/3)	[2/1]	* tetragonal,	235	509	29)
		Na	GTO	(10/3)	[3/1]	* part. optimized	Na—N		
Na^+	5 NH_3	H	GTO	(4)	[1]	STO-3G geometry			
		N	GTO	(7/3)	[2/1]	* pentagonal,	230	585	29)
		Na	GTO	(10/3)	[3/1]	* part. optimized	Na—N		
Na^+	6 NH_3	H	GTO	(4)	[1]	STO-3G geometry			
		N	GTO	(7/3)	[2/1]	* hexagonal,	230	628	29)
		Na	GTO	(10/3)	[3/1]	* part. optimized	Na—N		
Na^+	H_2O	H, O	GTO	STO-3G		experimental	200	178	31)
		Na	GTO	STO-3G + 2p orbital			Na—O		
Na^+	H_2O	H, O	GTO	STO-3G		experimental	200	126	31)
		Na	GTO	STO-3G reoptimized			Na—O		
Na^+	H_2O	H	GTO	(3)	[1]				
		O	GTO	(7/3)	[2/1]	experimental	215	121	31)
		Na	GTO	STO-3G reoptimized			Na—O		
Na^+	H_2O	H, O	GTO	4-31G		experimental	200	175	31)
		Na	GTO	STO-3G reoptimized			Na—O		
Na^+	H_2O	H, O	GTO	4-31G			220	138	
		Na	GTO	STO-3G + 2p orbital, reoptimized		experimental	Na—O		31)
Na^+	H_2O	H	GTO	(6/2)	[2/2]				
		O	GTO	(11/7/2)	[4/3/2]	not specified	225	100	33)
		Na	GTO	(13/8/2)	[7/4/2]		Na—O		
Na^+	H_2O	H	GTO	(6/1)	[2/1]				
		O	GTO	(11/7/1)	[4/3/1]	not specified	225	105	76)
		Na	GTO	(13/8/1)	[7/4/1]		Na—O		
Na^+	H_2O	H	GTO	(6/1)	[3/1]				
		O	GTO	(11/7/1)	[5/4/1]	experimental	224	105	34)
		Na	GTO	(14/8/1)	[8/6/1]		Na—O		
Na^+	H_2O	H	GTO	(4/1)	[2/1]				
		O	GTO	(9/5/1)	[4/2/1]	not specified	220	113	36)
		Na	GTO	(11/7)	[6/4]		Na—O		
Na^+	H_2O	all	GTO	3-21G		* fully optimized	212	173	19)
							Na—O		

Table 1. (continued)

Ion	Ligand	Basis set[b] atom	type	uncontracted	contracted	Ligand geometry/ * Complex geometry[c]	R [pm]	—E(Int) [kJ/mole]	Ref.
Na^+	H_2O	all	GTO	6-31G*		* 3-21G geometry	212 Na—O	118	[19]
Na^+	H_2O	H	GTO	(6/1)	[3/1]				
		O	GTO	(11/7/1)	[5/4/1]	experimental	220	104	[20]
		Na	GTO	(12/6)	[6/4]		Na—O		
Na^+	H_2O	H, O	GTO	STO-3G		not specified	199	170	[77]
		Na	GTO	STO-3G reoptimized			Na—O		
Na^+	H_2O	H	GTO	(6/1)	[2/1]				
		O	GTO	(12/8/2)	[5/8/2]	experimental	225	102	[78]
		Na	GTO	(13/8/1)	[7/4/1]		Na—O		
Na^+	$4\ H_2O$	H	GTO	(4)	[1]	STO-3G geometry			
		O	GTO	(7/3)	[2/1]	* tetragonal,	210	477	[29]
		Na	GTO	(10/3)	[3/1]	* part. optimized	Na—O		
Na^+	$5\ H_2O$	H	GTO	(4)	[1]	STO-3G geometry			
		O	GTO	(7/3)	[2/1]	* pentagonal,	220	552	[29]
		Na	GTO	(10/3)	[3/1]	* part. optimized	Na—O		
Na^+	$6\ H_2O$	H	GTO	(4)	[1]	STO-3G geometry			
		O	GTO	(7/3)	[2/1]	* hexagonal,	220	594	[29]
		Na	GTO	(10/3)	[3/1]	* part. optimized	Na—O		
Na^+	HF	all	GTO	3-21G		* fully optimized	203 Na—F	131	[19]
Na^+	HF	all	GTO	6-31G*		* 3-21G geometry	203 Na—F	87	[19]
Na^+	HCN	all	GTO	3-21G		* fully optimized	224 Na—N	145	[19]
Na^+	HCN	all	GTO	6-31G*		* 3-21G geometry	224 Na—N	115	[19]
Na^+	H_2CO	all	GTO	3-21G		* fully optimized	208 Na—O	159	[19]

Na^+	H_2CO	all	GTO 6-31G*		* 3-21G geometry	208 Na—O	121	[19]
Na^+	H_2CO	all	GTO STO-3G		* fully optimized	not given	141	[48]
Na^+	$4\ H_2CO$	H C, O Na	GTO (4) GTO (7/3) GTO (10/3)	[1] [2/1] [3/1]	STO-3G geometry * tetragonal, * part. optimized	210 Na—O	387	[29]
Na^+	$5\ H_2CO$	H C, O Na	GTO (4) GTO (7/3) GTO (10/3)	[1] [2/1] [3/1]	STO-3G geometry * pentagonal, * part. optimized	220 Na—O	441	[29]
Na^+	$6\ H_2CO$	H C, O Na	GTO (4) GTO (7/3) GTO (10/3)	[1] [2/1] [3/1]	STO-3G geometry * hexagonal, * part. optimized	230 Na—O	464	[29]
Na^+	$CH_3—NH_2$	all	GTO 3-21G		* fully optimized	229 Na—N	169	[19]
Na^+	$CH_3—NH_2$	all	GTO 6-31G*		* 3-21G geometry	229 Na—N	132	[19]
Na^+	$CH_3—OH$	all	GTO 3-21G		* fully optimized	211 Na—O	172	[19]
Na^+	$CH_3—OH$	all	GTO 6-31G*		* 3-21G geometry	211 Na—O	121	[19]
Na^+	$CH_3—F$	all	GTO 3-21G		* fully optimized	202 Na—F	144	[19]
Na^+	$CH_3—F$	all	GTO 6-31G*		* 3-21G geometry	202 Na—F	105	[19]
Na^+	PH_3	all	GTO 3-21G		* fully optimized	294 Na—P	86	[19]
Na^+	PH_3	all	GTO 6-31G*		* 3-21G geometry	294 Na—P	74	[19]
Na^+	H_2S	all	GTO 3-21G		* fully optimized	278 Na—S	74	[19]
Na^+	H_2S	all	GTO 6-31G*		* 3-21G geometry	278 Na—S	52	[19]
Na^+	HCl	all	GTO 3-21G		* fully optimized	279 Na—Cl	51	[19]

Table 1. (continued)

Ion	Ligand	Basis set[b] atom	type	uncontracted	contracted	Ligand geometry/ * Complex geometry[c]	R [pm]	—E(Int) [kJ/mole]	Ref.
Na^+	HCl	all	GTO	6-31G*		* 3-21G geometry	279 Na—Cl	37	19)
Na^+	CH_3—CN	all	GTO	3-21G		* fully optimized	221 Na—N	170	19)
Na^+	CH_3—CN	all	GTO	6-31G*		* 3-21G geometry	221 Na—N	140	19)
Na^+	CH_2–NH–CH_2 (ring)	all	GTO	6-21G		6-21G geometry of Li complex	225 Na—N	178	56)
Na^+	CH_2–O–CH_2 (ring)	all	GTO	6-21G		6-21G geometry of Li complex	218 Na—N	172	56)
Na^+	CH_3—CHO	all	GTO	3-21G		* fully optimized	206 Na—O	176	19)
Na^+	CH_3—CHO	all	GTO	6-31G*		* 3-21G geometry	206 Na—O	138	19)
Na^+	CH_3—NH—CH_3	all	GTO	3-21G		* fully optimized	228 Na—N	163	19)
Na^+	CH_3—NH—CH_3	all	GTO	6-31G*		* 3-21G geometry	228 Na—N	125	19)
Na^+	HCO—NH_2	H C, N, O Na	GLO GLO GLO	(1) (2/1) (4/2)	[1] [2/1] [2/1]	experimental	not given	131	79)
Na^+	CH_3—O—CH_3	all	GTO	3-21G		* fully optimized	210 Na—O	169	19)
Na^+	CH_3—O—CH_3	all	GTO	6-31G*		* 3-21G geometry	210 Na—O	118	19)
Na^+	CH_3—O—CH_3	H, C, O Na	GTO GTO	STO-3G STO-3G reoptimized		experimental	200 Na—O	163	61)
Na^+	CH_3—O—CH_3	H C, O Na	GTO GTO GTO	(3) (7/3) (9/6)	[1] [2/1] [3/2]	experimental	202 Na—O	90	62)

Na^+	CH_3—PH_2	all	GTO 3-21G		* fully optimized	291 Na—P	98	19)
Na^+	CH_3—PH_2	all	GTO 6-31G*		* 3-21G geometry	291 Na—P	92	19)
Na^+	CH_3—SH	all	GTO 3-21G		* fully optimized	273 Na—S	90	19)
Na^+	CH_3—SH	all	GTO 6-31G*		* 3-21G geometry	273 Na—S	71	19)
Na^+	CH_3—Cl	all	GTO 3-21G		* fully optimized	264 Na—Cl	84	19)
Na^+	CH_3—Cl	all	GTO 6-31G*		* 3-21G geometry	264 Na—Cl	58	19)
Na^+	HC=O | HC=O	H C, O Na	GLO (4) GLO (8/4) GLO (11/7)	[3] [4/2] [6/4]	experimental	not given	158	64)
Na^+	HC=O | HC=O	H C, O Na	GLO (1) GLO (2/1) GLO (4/2)	[1] [2/1] [2/1]	experimental	not given	110	65)
Na^+	CH_3—CO—NH_2	H C, N, O Na	GLO (1) GLO (2/1) GLO (4/2)	[1] [2/1] [2/1]	experimental	209 Na—O	142	67)
Na^+	CH_3—PH—CH_3	all	GTO 3-21G		* fully optimized	288 Na—P	108	19)
Na^+	CH_3—PH—CH_3	all	GTO 6-31G*		* 3-21G geometry	288 Na—P	105	19)
Na^+	CH_3—S—CH_3	all	GTO 3-21G		* fully optimized	269 Na—S	101	19)
Na^+	CH_3—S—CH_3	all	GTO 6-31G*		* 3-21G geometry	269 Na—S	84	19)
Na^+	CH_3—S—CH_3	H C S Na	GTO (3) GTO (7/3) GTO (9/6) GTO (7/3)	[1] [2/1] [3/2] [2/1]	assumed	254 Na—S	57	62)
Na^+	CHO—CH_2—CHO	H C, O Na	GLO (1) GLO (2/1) GLO (4/2)	[1] [2/1] [2/1]	assumed	not given	175	65)

Table 1. (continued)

Ion	Ligand	Basis set[b] atom	type	uncontracted	contracted	Ligand geometry/ * Complex geometry[c]	R [pm]	—E(Int) [kJ/mole]	Ref.
Na^+	CH_3—CH_2—CO—NH_2	H	GLO	(1)	[1]				
		C, N, O	GLO	(2/1)	[2/1]	experimental	209	149	[67]
		Na	GLO	(4/2)	[2/1]		Na—O		
Na^+	CH_3—CO—NH—CH_3	Ligand	GTO	STO-3G		part. optimized	195	208	[77]
		Na	GTO	STO-3G reoptimized			Na—O		
Na^+	CH_3—CH_2—O—CH_2—CH_3	H	GTO	(3)	[1]				
		C, O	GTO	(7/3)	[2/1]	experimental	225	93	[62]
		Na	GTO	(7/3)	[2/1]		Na—O		
Na^+	CH_3—COO—CH_3	Ligand	GTO	STO-3G		part. optimized	199	162	[77]
		Na	GTO	STO-3G reoptimized			Na—O		
Na^+	CHO—CH_2—CH_2—CHO	H	GLO	(1)	[1]				
		C, O	GLO	(2/1)	[2/1]	assumed	not	196	[65]
		Na	GLO	(4/2)	[2/1]		given		
Na^+	CH_3—CH_2—CH_2—CO—NH_2	H	GLO	(1)	[1]				
		C, N, O	GLO	(2/1)	[2/1]	experimental	208	153	[67]
		Na	GLO	(4/2)	[2/1]		Na—O		
Na^+	CH_3—O—CH_2—CH_2—O—CH_3	H	GTO	(3)	[1]	assumed			
		C, O	GTO	(7/3)	[2/1]	synperiplanar	212	172	[70]
		Na	GTO	(7/3)	[2/1]	E(conf) = 30 kJ/mole	Na—O		
Na^+	NH_2—CO—NH—CO—NH_2	H	GLO	(1)	[1]				
		C, N, O	GLO	(2/1)	[2/1]	experimental	210	247	[71]
		Na	GLO	(4/2)	[2/1]		Na—O		
Na^+	2 NH_2—CO—NH—CO—NH_2	H	GLO	(1)	[1]				
		C, N, O	GLO	(2/1)	[2/1]	experimental	not	417	[71]
		Na	GLO	(4/2)	[2/1]		given		
Na^+	CO—N=C(NH_2)—CH=CH—NH (ring) cytosine	H	GTO	(3)	[1]	assumed	215		
		C, N, O	GTO	(7/3)	[2/1]	* not optimized	Na—O	216	[80]
		Na	GTO	STO-3G reoptimized			230		
							Na—N		

Na^+	CO–N=C(NH_2)–CH=CH–NH cytosine	H C, N, O Na	GLO (1) GLO (2/1) GLO (4/2) counterpoise corrected	[1] [2/1] [2/1]	experimental	220 Na—O	191	73)
Na^+	CO–NH–CH=CH–CO–NH uracil	H C, N, O Na	GTO (3) GTO (7/3) GTO STO-3G reoptimized	[1] [2/1]	assumed * not optimized	200 Na—O	138	80)
Na^+	CH_3–CO–N(CH_2–CH_3)$_2$	H C, N, O Na	GTO (3) GTO (7/3) GTO (7/3)	[1] [2/1] [2/1]	assumed	201 Na—O	139	62)
Na^+	CO–NH–CH=C(CH_3)–CO–NH thymine	H C, N, O Na	GLO (1) GLO (2/1) GLO (4/2) counterpoise corrected	[1] [2/1] [2/1]	experimental	210 Na—O	122	73)
Na^+	O––––––––CH_2 HO–CH_2–CH–CH(OH)–CH–OH ribose	H C, O Na	GTO (3) GTO (7/3) not specified	[1] [2/1]	experimental	not given	146	81)
Na^+	N–C(NH_2)=C–N=CH CH––N==C–––NH adenine	H C, N Na	GTO (3) GTO (7/3) GTO STO-3G reoptimized	[1] [2/1]	assumed * not optimized	215 Na—O	110	80)
Na^+	N–C(NH_2)=C–N=CH CH––N==C–––NH adenine	H C, N Na	GLO (1) GLO (2/1) GLO (4/2) counterpoise corrected	[1] [2/1] [2/1]	experimental	220 Na—N	132	73)
Na^+	NH–CO–C–N=CH NH_2–C==N––C–––NH guanine	H C, N, O Na	GTO (3) GTO (7/3) GTO STO-3G reoptimized	[1] [2/1]	assumed * not optimized	215 Na—O 290 Na—N	225	80)
Na^+	NH–CO–C–N=CH NH_2–C==N––C–––NH guanine	H C, N, O Na	GLO (1) GLO (2/1) GLO (4/2) counterpoise corrected	[1] [2/1] [2/1]	experimental	225 Na—N	210	73)
Na^+	CH_2–NH–CH_2–CH_2–NH–CH_2 CH_2–NH–CH_2–CH_2–NH–CH_2	H C, N Na	GLO (1) GLO (2/1) GLO (4/2)	[1] [2/1] [2/1]	assumed * part. optimized	208 Na—N	182	74)

Table 1. (continued)

Ion	Ligand	Basis set[b] atom	type	uncontracted	contracted	Ligand geometry/ * Complex geometry[c]	R [pm]	—E(Int) [kJ/mole]	Ref.
Na^+	$CH_3-NH-CH_2-CH_2-NH-CH_2$	H	GLO	(1)	[1]	assumed			
	$CH_3-NH-CH_2-CH_2-NH-CH_2$	C, N	GLO	(2/1)	[2/1]	* not varied	208	79	74)
		Na	GLO	(4/2)	[2/1]		Na—N		
Na^+	$CH_2-O-CH_2-CH_2-O-CH_2$	H	GLO	(1)	[1]	assumed			
	$CH_2-O-CH_2-CH_2-O-CH_2$	C, O	GLO	(2/1)	[2/1]	* part. optimized	228	247	74)
		Na	GLO	(4/2)	[2/1]		Na—O		
Na^+	$CH_3-O-CH_2-CH_2-O-CH_2$	H	GLO	(1)	[1]	assumed			
	$CH_3-O-CH_2-CH_2-O-CH_2$	C, O	GLO	(2/1)	[2/1]	* not varied	228	42	74)
		Na	GLO	(4/2)	[2/1]		Na—O		
Na^+	$CH_2-(O-CH_2-CH_2)_2-O-CH_2$	H	GTO	(3)	[1]	geometry of			
	$CH_2-(O-CH_2-CH_2)_2-O-CH_2$	C, O	GTO	(7/3)	[2/1]	complexed ligand	245	343	75)
		Na	GTO	(7/4)	[2/1]	* experimental	Na—O		
K^+	NH_3	H, N	GTO	4-31G		assumed	278	101	82)
		K	GTO	(14/9)	[8/6]		K—N		
K^+	NH_3	H	GTO	(6/1)	[3/1]				
		N	GTO	(11/7/1)	[5/4/1]	experimental	290	77	20)
		K	GTO	(14/9/5)	[8/4/2]		K—N		
K^+	NH_3	H	GTO	(4/1)	[3/1]				
		N	GTO	(9/5/1)	[5/3/1]	experimental	290	81	21)
		K	GTO	(14/9)	[8/4]		K—N		
K^+	NH_3	H	GTO	(6/1)	[3/1]				
		N	GTO	(11/7/1)	[5/4/1]	experimental	290	77	21)
		K	GTO	(14/9)	[8/4]		K—N		
K^+	$4 NH_3$	H	GTO	(4)	[1]	STO-3G geometry			
		N	GTO	(7/3)	[2/1]	* tetragonal,	270	333	29)
		K	GTO	(13/6)	[4/2]	* part. optimized	K—N		
K^+	$5 NH_3$	H	GTO	(4)	[1]	STO-3G geometry			
		N	GTO	(7/3)	[2/1]	* pentagonal,	280	403	29)
		K	GTO	(13/6)	[4/2]	* part. optimized	K—N		

K^+	6 NH_3	H	GTO	(4)	[1]	STO-3G geometry			
		N	GTO	(7/3)	[2/1]	* hexagonal,	285	457	29)
		K	GTO	(13/6)	[4/2]	* part. optimized	K—N		
K^+	H_2O	H, O	GTO	STO-3G		experimental	250	82	31)
		K	GTO	STO-3G reoptimized			K—O		
K^+	H_2O	H	GTO	(3)	[1]				
		O	GTO	(7/3)	[2/1]	experimental	257	85	31)
		K	STO-3G reoptimized				K—O		
K^+	H_2O	H, O	GTO	4-31G		experimental	250	115	31)
		K	GTO	STO-3G reoptimized			K—O		
K^+	H_2O	H, O	GTO	4-31G			260	101	
		K	GTO	STO-3G + 3p orbital, reoptimized		experimental	K—O		31)
K^+	H_2O	H	GTO	(6/2)	[2/2]				
		O	GTO	(11/7/2)	[4/3/2]	not specified	269	70	33)
		K	GTO	(17/11/2)	[9/6/2]		K—O		
K^+	H_2O	H	GTO	(6/1)	[2/1]				
		O	GTO	(11/7/1)	[4/3/1]	not specified	269	73	76)
		K	GTO	(17/11/1)	[11/7/1]		K—O		
K^+	H_2O	H	GTO	(4/1)	[2/1]				
		O	GTO	(9/5/1)	[4/2/1]	not specified	265	75	36)
		K	GTO	(14/9)	[8/6]		K—O		
K^+	H_2O	H, O	GTO	4-31G		assumed	259	98	82)
		K	GTO	(14/9)	[8/6]		K—O		
K^+	H_2O	H	GTO	(6/1)	[3/1]				
		O	GTO	(11/7/1)	[5/4/1]	experimental	270	72	20)
		K	GTO	(14/9/5)	[8/4/2]		K—O		
K^+	H_2O	H	GTO	(4/1)	[3/1]				
		O	GTO	(9/5/1)	[5/3/1]	experimental	270	76	21)
		K	GTO	(14/9)	[8/4]		K—O		
K^+	H_2O	H	GTO	(6/1)	[3/1]				
		O	GTO	(11/7/1)	[5/4/1]	experimental	270	72	21)
		K	GTO	(14/9)	[8/4]		K—O		
K^+	H_2O	H, O	GTO	STO-3G		not specified	240	117	77)
		K	GTO	STO-3G reoptimized			K—O		

Table 1. (continued)

Ion	Ligand	Basis set[b] atom	type	uncontracted	contracted	Ligand geometry/ * Complex geometry[c]	R [pm]	−E(Int) [kJ/mole]	Ref.
K^+	$4\,H_2O$	H	GTO	(4)	[1]	STO-3G geometry			
		O	GTO	(7/3)	[2/1]	* tetragonal,	260	320	29)
		K	GTO	(13/6)	[4/2]	* part. optimized	K—O		
K^+	$5\,H_2O$	H	GTO	(4)	[1]	STO-3G geometry			
		O	GTO	(7/3)	[2/1]	* pentagonal,	265	387	29)
		K	GTO	(13/6)	[4/2]	* part. optimized	K—O		
K^+	$6\,H_2O$	H	GTO	(4)	[1]	STO-3G geometry			
		O	GTO	(7/3)	[2/1]	* hexagonal,	265	441	29)
		K	GTO	(13/6)	[4/2]	* part. optimized	K—O		
K^+	HCN	H, C, N	GTO	4-31G		assumed	270	88	82)
		K	GTO	(14/9)	[8/6]		K—N		
K^+	$4\,H_2CO$	H	GTO	(4)	[1]	STO-3G geometry			
		C, O	GTO	(7/3)	[2/1]	* tetragonal,	270	265	29)
		K	GTO	(13/6)	[4/2]	* part. optimized	K—O		
K^+	$5\,H_2CO$	H	GTO	(4)	[1]	STO-3G geometry			
		C, O	GTO	(7/3)	[2/1]	* pentagonal,	265	316	29)
		K	GTO	(13/6)	[4/2]	* part. optimized	K—O		
K^+	$6\,H_2CO$	H	GTO	(4)	[1]	STO-3G geometry			
		C, O	GTO	(7/3)	[2/1]	* hexagonal,	270	354	29)
		K	GTO	(13/6)	[4/2]	* part. optimized	K—O		
K^+	$CH_3—NH_2$	H, C, N	GTO	4-31G		assumed	278	96	82)
		K	GTO	(14/9)	[8/6]		K—N		
K^+	$CH_3—NH_2$	H	GTO	(4/1)	[3/1]				
		C, N	GTO	(9/5/1)	[5/3/1]	experimental	290	79	21)
		K	GTO	(14/9)	[8/4]		K—N		
K^+	$CH_3—NH_2$	H	GTO	(6/1)	[3/1]	experimental			
		C, N	GTO	(11/7/1)	[5/4/1]	* not varied	290	76	21)
		K	GTO	(14/9)	[8/4]		K—N		
K^+	$CH_3—OH$	H, C, O	GTO	4-31G		assumed	259	99	82)
		K	GTO	(14/9)	[8/6]	* not varied	K—O		

K^+	H_2S	H, S	GTO	4-31G		experimental	330	49	55)
		K	GTO	(14/9)	[8/6]		K—S		
K^+	CH_3—CN	H, C, N	GTO	4-31G		assumed	273	107	82)
		K	GTO	(14/9)	[8/6]	* not varied	K—N		
K^+	CH_3—NH—CH_3	H, C, N	GTO	4-31G		assumed	278	79	82)
		K	GTO	(14/9)	[8/6]	* not varied	K—N		
K^+	HCO—NH_2	Ligand	GTO	4-31G		assumed	259	124	82)
		K	GTO	(14/9)	[8/6]	* not varied	K—O		
K^+	CH_3—O—CH_3	H, C, O	GTO	4-31G		assumed	259	97	82)
		K	GTO	(14/9)	[8/6]	* not varied	K—O		
K^+	CH_3—O—CH_3	H, C, O	GTO	STO-3G		experimental	240	108	61)
		K	GTO	STO-3G reoptimized			K—O		
K^+	HC=O	H	GLO	(1)	[1]				
	| HC=O	C, O	GLO	(2/1)	[2/1]	experimental	not	146	64)
		K	GLO	(8/4)	[3/2]		given		
K^+	CH_3—CO—NH_2	H	GLO	(1)	[1]	experimental			
		C, N, O	GLO	(2/1)	[2/1]	* not varied	250	131	67)
		K	GLO	(8/4)	[3/2]		K—O		
K^+	CHO—CH_2—CHO	H	GLO	(1)	[1]				
		C, O	GLO	(2/1)	[2/1]	assumed	not	169	65)
		K	GLO	(8/4)	[3/2]		given		
K^+	CH_3—CO—NH—CH_3	Ligand	GTO	STO-3G		part. optimized	235	147	77)
		K	GTO	STO-3G reoptimized			K—O		
K^+	CH_3—COO—CH_3	Ligand	GTO	STO-3G		part. optimized	240	107	77)
		K	GTO	STO-3G reoptimized			K—O		
K^+	CHO—CH_2—CH_2—CHO	H	GLO	(1)	[1]				
		C, O	GLO	(2/1)	[2/1]	assumed	not	187	65)
		K	GLO	(8/4)	[3/2]		given		
K^+	NH_2—CO—NH—CO—NH_2	H	GLO	(1)	[1]				
		C, N, O	GLO	(2/1)	[2/1]	experimental	250	249	71)
		K	GLO	(8/4)	[3/2]		K—O		
K^+	CH_2–(O–CH_2–CH_2)$_2$–O–CH_2	H	GTO	(3)	[1]	geometry of			
	| CH_2–(O–CH_2–CH_2)$_2$–O–CH_2 |	C, O	GTO	(7/3)	[2/1]	complexed ligand	277	263	75)
		K	GTO	(9/6)	[3/2]	* experimental	K—O		
NH_4^+	NH_3	all	GTO	STO-3G		assumed	250	177	83)
							N—N		

Table 1. (continued)

Ion	Ligand	Basis set[b] atom	type	uncontracted	contracted	Ligand geometry/ * Complex geometry[c]	R [pm]	—E(Int) [kJ/mole]	Ref.
NH_4^+	NH_3	all	GTO	4-31G		assumed	275 N—N	131	[83]
NH_4^+	NH_3	H	GTO	(3/1)	[2/1]	* part. optimized	278 N—N	118	[85]
		N	GTO	(8/4/1)	[4/2/1]				
NH_4^+	NH_3	H	GLO	(5)?	[1]				
		H7	GLO	(5/1)?	[2/1]	* part. optimized	271 N—N	151	[86]
		N	GLO	(10/5)	[4/2]				
		H7 is involved in hydrogen bond							
NH_4^+	NH_3	all	GTO	4-31G		4-31G geometry	279 N—N	112	[87]
NH_4^+	$2\ NH_3$	all	GTO	STO-3G		assumed	260 N—N	310	[84]
NH_4^+	$3\ NH_3$	all	GTO	STO-3G		assumed	265 N—N	407	[84]
NH_4^+	$4\ NH_3$	all	GTO	STO-3G		assumed	270 N—N	481	[84]
NH_4^+	$4\ NH_3$	all	GTO	STO-3G		assumed	270 N—N	475	[83]
NH_4^+	$5\ NH_3$	all	GTO	STO-3G		assumed	290 N—N	524	[84
NH_4^+	H_2O	H	GTO	(3)	[1]	experimental	270 N—O	95	[31]
		N, O	GTO	(7/3)	[2/1]				
NH_4^+	H_2O	all	GTO	STO-3G		experimental	240 N—O	156	[83] [84]
NH_4^+	H_2O	all	GTO	4-31G		experimental	265 N—O	118	[83]
NH_4^+	H_2O	all	GTO	3-21G		* fully optimized	259 N—O	142	[88]
NH_4^+	H_2O	all	GTO	6-31G*		* 3-21G geometry	259 N—O	82	[88]

NH_4^+	H_2O	all	GTO 6-31G*		*	part. optimized	not given	90	[88]
NH_4^+	H_2O	all	GTO 4-31G			experimental	268 N—O	114	[44]
NH_4^+	H_2O	all	GTO 4-31G			experimental	268 N—O	114	[38]
NH_4^+	H_2O	H N, O	GTO (3/1) GTO (8/4/1)	[2/1] [4/2/1]	*	part. optimized	271 N—O	98	[85]
NH_4^+	H_2O	H N, O	GTO (4/1) GTO (9/5/1)	[2/1] [3/2/1]		assumed	278 N—O	86	[89]
NH_4^+	H_2O	H N, O counterpoise corrected	GTO (4/1) GTO (9/5/1)	[2/1] [3/2/1]		assumed	279 N—O	84	[89]
NH_4^+	H_2O	H N, O	GTO (5/1) GTO (10/6/1)	[3/1] [5/3/1]		assumed	278 N—O	83	[89]
NH_4^+	H_2O	H N, O	GTO (4/1) GTO (9/5/2)	[2/1] [3/2/2]		assumed	279 N—O	82	[89]
NH_4^+	H_2O	H N, O	GTO (6/1) GTO (11/7/2)	[2/1] [4/3/2]		experimental	278 N—O	77	[90]
NH_4^+	H_2O	H N, O counterpoise corrected	GTO (6/1) GTO (11/7/2)	[2/1] [4/3/2]		experimental	278 N—O	77	[90]
NH_4^+	H_2O	H N, O	GTO (4/1) GTO (9/5/1)	[2/1] [4/2/1]	*	fully optimized	276 N—O	86	[91]
NH_4^+	H_2O	H N, O	GTO (4/1) GTO (9/5/1)	[2/1] [4/2/1]		isolated partners fully optimized	277 N—O	85	[91]
NH_4^+	H_2O	H N, O	GTO (4/1) GTO (9/5/1)	[2/1] [4/2/1]		experimental	278 N—O	87	[91]
NH_4^+	H_2O	H N, O counterpoise corrected	GTO (4/1) GTO (9/5/1)	[2/1] [4/2/1]		experimental	278 N—O	85	[91]
NH_4^+	H_2O	H N, O	GTO (4) GTO (9/5)	[2] [4/2]	* *	same geometry as 2 entries above	278 N—O	108	[91]
NH_4^+	H_2O	H N, O counterpoise corrected	GTO (4) GTO (9/5)	[2] [4/2]	* *	same geometry as 3 entries above	278 N—O	105	[91]

Table 1. (continued)

Ion	Ligand	Basis set[b] atom	type	uncontracted	contracted	Ligand geometry/ * Complex geometry[c]	R [pm]	—E(Int) [kJ/mole]	Ref.
NH_4^+	H_2O	H N, O	GTO GTO	(4/1) (9/5/2)	[2/1] [4/2/1]	* same geometry as * 4 entries above	278 N—O	81	91)
NH_4^+	H_2O	H N, O	GTO GTO	(4/1) (9/5/2)	[2/1] [4/3/2]	* same geometry as * 5 entries above	278 N—O	81	91)
NH_4^+	H_2O	H N, O	GTO GTO	(5/1) (10/6/2)	[3/1] [5/4/2]	* same geometry as * 6 entries above	278 N—O	78	91)
NH_4^+	2 H_2O	all	GTO	STO-3G		experimental	245 N—O	275	83) 84)
NH_4^+	2 H_2O	all	GTO	3-21G		* fully optimized	266 N—O	255	88)
NH_4^+	2 H_2O	all	GTO	6-31G*		* 3-21G geometry	259 N—O	155	88)
NH_4^+	2 H_2O	all	GTO	6-31G*		* part.optimized	not given	166	88)
NH_4^+	3 H_2O	all	GTO	3-21G		* fully optimized	272 N—O	346	88)
NH_4^+	3 H_2O	all	GTO	6-31G*		* 3-21G geometry	259 N—O	217	88)
NH_4^+	3 H_2O	all	GTO	STO-3G		experimental	255 N—O	367	83) 84)
NH_4^+	4 H_2O	all	GTO	STO-3G		experimental	260 N—O	441	83) 84)
NH_4^+	4 H_2O	all	GTO	3-21G		* fully optimized	277 N—O	427	88)
NH_4^+	4 H_2O	all	GTO	6-31G*		* 3-21G geometry	259 N—O	276	88)
NH_4^+	5 H_2O	all	GTO	STO-3G		experimental	262 N—O	493	84)
NH_4^+	6 H_2O	all	GTO	4-31G		experimental * not varied	268 N—O	659	38)

NH_4^+	CH_3-O-CH_3	all	GTO STO-3G		not specified	not given	161	[92]
NH_4^+	HCOOH	all	GTO STO-3G		experimental	244 N—O	122	[93]
NH_4^+	$HO-CH_2-CH_2-OH$	all	GTO STO-3G		not specified	not given	198	[92]
Be^{++}	NH_3	all	GTO 4-31G		assumed	164 Be—N	686	[44]
Be^{++}	H_2O	H O Be	GTO (4/1) GTO (9/5/1) GTO (9/3)	[2/1] [4/2/1] [4/2]	not specified	150 Be—O	586	[36]
Be^{++}	H_2O	H, O Be	GTO 4-31G not specified		* fully optimized	156 Be—O	not given	[94]
Be^{++}	H_2O	all	GTO 4-31G		experimental	159 Be—O	601	[44]
Be^{++}	H_2O	H O Be	GTO (4) GTO (9/10) GTO (9/2)	[4] [6/4] [7/1]	experimental	156 Be—O	592	[39]
Be^{++}	H_2O	H O Be	GTO (6/1) GTO (11/7/1) GTO (8)	[2/1] [4/3/1] [2]	not specified	159 Be—O	467	[95]
Be^{++}	H_2O	H O Be	GTO (6/1) GTO (11/7/1) GTO (11/3)	[2/1] [4/3/1] [4/2]	not specified	153 Be—O	563	[95]
Be^{++}	$HCO-NH_2$	all	GTO 4-31G		experimental	148 Be—N	824	[44]
Be^{++}	$NCO-NH_2$	H C, O Be	GLO (1) GLO (2/1) GLO (2)	[1] [2/1] [1]	experimental	not given	594	[79]
Be^{++}	HC=O | HC=O	H C, O Be	GLO (4) GLO (8/4) GLO (8)	[3] [4/2] [4]	experimental	not given	634	[64]
Be^{++}	2 HC=O | HC=O	H C, O Be	GTO (4) GTO (7/3) GTO (7/3)	[2] [4/2] [4/1]	assumed * part. optimized	185 Be—O	1350	[66]

Table 1. (continued)

Ion	Ligand	Basis set[b] atom	type	uncontracted	contracted	Ligand geometry/ * Complex geometry[c]	R [pm]	−E(Int) [kJ/mole]	Ref.
Be^{++}	$N(CH_3)_3$	all	GTO	4-31G		* not optimized	164 Be—N	799	44)
Be^{++}	CH_3—CO—NH_2	H	GLO	(1)	[1]				
		C, N, O	GLO	(2/1)	[2/1]	experimental	145	649	67)
		Be	GLO	(2)	[1]		Be—O		
Be^{++}	CH_3—CH_2—CO—NH_2	H	GLO	(1)	[1]				
		C, N, O	GLO	(2/1)	[2/1]	experimental	145	676	67)
		Be	GLO	(2)	[1]		Be—O		
Be^{++}	CH_3—CH_2—CH_2—CO—NH_2	H	GLO	(1)	[1]				
		C, N, O	GLO	(2/1)	[2/1]	experimental	144	691	67)
		Be	GLO	(2)	[1]		Be—O		
Be^{++}	2 HC=S—HC=S	H	GTO	(4)	[2]				
		C	GTO	(7/3)	[4/2]	assumed	210	1110	66)
		S	GTO	(10/6)	[5/4]	* part. optimized	Be—O		
		Be	GTO	(7/3)	[4/1]				
Be^{++}	NH_2—CO—NH—CO—NH_2	H	GLO	(1)	[1]				
		C, N, O	GLO	(2/1)	[2/1]	experimental	172	795	71)
		Be	GLO	(2)	[1]		Be—O		
Be^{++}	2 NH_2—CO—NH—CO—NH_2	H	GLO	(1)	[1]				
		C, N, O	GLO	(2/1)	[2/1]	experimental	not given	967	71)
		Be	GLO	(2)	[1]				
Mg^{++}	H_2O	H	GTO	(4)	[2]				
		O	GTO	(9/5)	[4/2]	assumed	192	355	96)
		Mg	GTO	(9/5)	[4/2]		Mg—O		
Mg^{++}	H_2O	H, O	GTO	STO-3G		experimental	180	490	31)
		Mg	GTO	STO-3G + 2p orbital			Mg—O		
Mg^{++}	H_2O	H, O	GTO	STO-3G		experimental	180	321	31)
		Mg	GTO	STO-3G reoptimized			Mg—O		

Mg^{++}	H_2O	H	GTO	(3)	[1]				
		O	GTO	(7/3)	[2/1]	experimental	187	317	[31]
		Mg	GTO	STO-3G reoptimized			Mg—O		
Mg^{++}	H_2O	H, O	GTO	4-31G		experimental	180	412	[31]
		Mg	GTO	STO-3G reoptimized			Mg—O		
Mg^{++}	H_2O	H, O	GTO	4-31G			190	387	
		Mg	GTO	STO-3G + 2p orbital, reoptimized		experimental	Mg—O		[31]
Mg^{++}	H_2O	H	GTO	(4/1)	[2/1]				
		O	GTO	(9/5/1)	[4/2/1]	not specified	195	335	[36]
		Mg	GTO	(11/7)	[6/4]		Mg—O		
Mg^{++}	H_2O	H	GTO	(4)	[2]				
		O	GTO	(7/3/1)	[4/2/1]	not specified	not	385	[97]
		Mg	GTO	(10/7)	[5/5]		given		
Mg^{++}	H_2O	H, O	GTO	4-31G		* fully optimized	189	not	[94]
		Mg	GTO	(10/5)	[6/4]		Mg—O	given	
Mg^{++}	6 H_2O	H, O	GTO	4-31G		assumed	207	1659	[94]
		Mg	GTO	(10/5)	[6/4]		Mg—O		
Mg^{++}	2 HC=NH—HC=NH	H	GTO	(4)	[2]		not		
		C, N	GTO	(7/3)	[4/2]	planar	given	1390	[97]
		Mg	GTO	(10/6)	[5/4]				
Mg^{++}	HC=O—HC=O	H	GLO	(4)	[3]				
		C, O	GLO	(8/4)	[4/2]	experimental	not	454	[64]
		Mg	GLO	(11/7)	[6/4]		given		
Mg^{++}	HC=O—HC=O	H	GLO	(1)	[1]				
		C, O	GLO	(2/1)	[2/1]	experimental	not	454	[65]
		Mg	GLO	(4/2)	[2/1]		given		
Mg^{++}	2 HC=O—HC=O	H	GTO	(4)	[2]		not		
		C, O	GTO	(7/3)	[4/2]	planar	given	955	[97]
		Mg	GTO	(10/6)	[5/4]				
Mg^{++}	CH_3—CO—NH_2	H	GLO	(1)	[1]	experimental			
		C, N, O	GLO	(2/1)	[2/1]	* not varied	185	444	[67]
		Mg	GLO	(4/2)	[2/1]		Mg—O		
Mg^{++}	HS—SH	H	GTO	(4)	[2]		not		
		S	GTO	(7/3/1)	[4/2/1]	* optimized?	given	317	[97]
		Mg	GTO	(10/6/1)	[5/4/1]				

Table 1. (continued)

Ion	Ligand	Basis set[b] atom	type	uncontracted	contracted	Ligand geometry/ * Complex geometry[c]	R [pm]	—E(Int) [kJ/mole]	Ref.
Mg^{++}	CHO—CH_2—CHO	H	GLO	(1)	[1]				
		C, O	GLO	(2/1)	[2/1]	assumed	not given	527	65)
		Mg	GLO	(4/2)	[2/1]				
Mg^{++}	CH_3—CH_2—CO—NH_2	H	GLO	(1)	[1]	experimental			
		C, N, O	GLO	(2/1)	[2/1]	* not varied	185 Mg—O	464	67)
		Mg	GLO	(4/2)	[2/1]				
Mg^{++}	CH_3—CO—NH—CH_3	H	GTO	(3)	[1]				
		C, N, O	GTO	(7/3)	[2/1]	assumed	175 Mg—O	414	98)
		Mg	GTO	(7/4)	[2/1]				
Mg^{++}	CHO—CH_2—CH_2—CHO	H	GLO	(1)	[1]				
		C, O	GLO	(2/1)	[2/1]	assumed	not given	607	65)
		Mg	GLO	(4/2)	[2/1]				
Mg^{++}	CH_3—CH_2—CH_2—CO—NH_2	H	GLO	(1)	[1]	experimental			
		C, N, O	GLO	(2/1)	[2/1]	* not varied	185 Mg—O	476	67)
		Mg	GLO	(4/2)	[2/1]				
Mg^{++}	CH_3—CO—$N(CH_3)_2$	H	GTO	(3)	[1]				
		C, N, O	GTO	(7/3)	[2/1]	assumed	177 Mg—O	421	98)
		Mg	GTO	(7/4)	[2/1]				
Mg^{++}	2 HC=S—HC=S	H	GTO	(4)	[2]				
		C	GTO	(7/3)	[4/2]	planar	not given	425	97)
		Mg, S	GTO	(10/6)	[5/4]				
Mg^{++}	NH_2—CO—CH_2—CO—NH_2	H	GTO	(3)	[1]				
		C, N, O	GTO	(7/3)	[2/1]	assumed	187 Mg—O	675	98)
		Mg	GTO	(7/4)	[2/1]				
Mg^{++}	NH_2—CO—NH—CO—NH_2	H	GLO	(1)	[1]				
		C, N, O	GLO	(2/1)	[2/1]	experimental	185 Mg—O	665	71)
		Mg	GLO	(4/2)	[2/1]				
Mg^{++}	2 NH_2—CO—NH—CO—NH_2	H	GLO	(1)	[1]				
		C, N, O	GLO	(2/1)	[2/1]	experimental	not given	1037	71)
		Mg	GLO	(4/2)	[2/1]				

Ion	Ligand	Atom	Basis			Geometry	Distance	Energy	Ref.
Mg^{++}	CO–N=C(NH_2)–CH=CH–NH (cytosine)	H	GLO	(1)	[1]				
		C, N, O	GLO	(2/1)	[2/1]	experimental	220	191	73)
		Mg	GLO	(4/2)	[2/1]		Mg—O		
		counterpoise corrected							
Mg^{++}	CO–NH–CH=CH–CO–NH (uracil)	H	GTO	(3)	[1]	experimental			
		C, N, O	GTO	(7/3)	[2/1]	* not optimized	175	434	99)
		Mg	GTO	STO-3G reoptimized			Mg—O		
Mg^{++}	CH_3—CH_2—CH_2—CO—N(CH_3)$_2$	H	GTO	(3)	[1]				
		C, N, O	GTO	(7/3)	[2/1]	assumed	174	431	98)
		Mg	GTO	(7/4)	[2/1]		Mg—O		
Mg^{++}	CH_3—CO—CH(OH)—S—CH_3	H	GTO	(3)	[1]	part. optimized;			
		C, N, O	GTO	(7/3)	[2/1]	* distances Mg—O	197	410	100)
		S	GTO	(10/6)	[3/2]	* optimized with	200		
		Mg	GTO	STO-3G reoptimized		* STO-3G basis set	Mg—O		
Mg^{++}	CO–NH–CH=C(CH_3)–CO–NH (thymine)	H	GLO	(1)	[1]				
		C, N, O	GLO	(2/1)	[2/1]	experimental	210	122	73)
		Mg	GLO	(4/2)	[2/1]		Mg—O		
		counterpoise corrected							
Mg^{++}	N–C(NH_2)=C–N=CH / CH––N==C–––NH (adenine)	H	GLO	(1)	[1]				
		C, N	GLO	(2/1)	[2/1]	experimental	220	132	73)
		Mg	GLO	(4/2)	[2/1]		Mg—N		
		counterpoise corrected							
Mg^{++}	NH–CO–C–N=CH / NH_2–C==N––C–––NH (guanine)	H	GLO	(1)	[1]				
		C, N, O	GLO	(2/1)	[2/1]	experimental	225	210	73)
		Mg	GLO	(4/2)	[2/1]		Mg—N		
		counterpoise corrected							
Mg^{++}	(CH_3)$_2$ N–CO–CH_2 / (CH_3)$_2$N–CO	H	GTO	(3)	[1]				
		C, N, O	GTO	(7/3)	[2/1]	assumed	182	691	98)
		Mg	GTO	(7/4)	[2/1]		Mg—O		
Mg^{++}	(CH_3)$_2$ N–CO–CH_2 / (CH_3)$_2$ N–CO–CH_2	H	GTO	(3)	[1]				
		C, N, O	GTO	(7/3)	[2/1]	assumed	180	705	98)
		Mg	GTO	(7/4)	[2/1]		Mg—O		
Ca^{++}	H_2O	H	GTO	(4)	[2]				
		O	GTO	(9/5)	[4/2]	assumed	236	242	96)
		Ca	GTO	(11/7)	[6/4]		Ca—O		

Table 1. (continued)

Ion	Ligand	Basis set[b] atom	type	uncontracted	contracted	Ligand geometry/ * Complex geometry[c]	R [pm]	—E(Int) [kJ/mole]	Ref.
Ca^{++}	H_2O	H, O	GTO	STO-3G		experimental	220	197	31)
		Ca	GTO	STO-3G reoptimized			Ca—O		
Ca^{++}	H_2O	H	GTO	(3)	[1]				
		O	GTO	(7/3)	[2/1]	experimental	230	213	31)
		Ca	GTO	STO-3G reoptimized			Ca—O		
Ca^{++}	H_2O	H, O	GTO	4-31G		experimental	230	274	31)
		Ca	GTO	STO-3G reoptimized			Ca—O		
Ca^{++}	H_2O	H, O	GTO	4-31G					
		Ca	GTO	STO-3G + 3p orbital, reoptimized		experimental	230 Ca—O	267	31)
Ca^{++}	H_2O	H	GTO	(4/1)	[2/1]				
		O	GTO	(9/5/1)	[4/2/1]	not specified	240	222	36)
		Ca	GTO	(14/9)	[8/6]		Ca—O		
Ca^{++}	H_2O	H, O	GTO	4-31G		* fully optimized	230	not	94)
		Ca	GTO	(12/7)	[8/6]		Ca—O	given	
Ca^{++}	6 H_2O	H, O	GTO	STO-3G		STO-3G geometry	230	1424	101)
		Ca	GTO	(14/9)	[6/3]		Ca—O		
Ca^{++}	6 H_2O	H, O	GTO	4-31G		assumed	240	1272	94)
		Ca	GTO	(12/7)	[8/6]		Ca—O		
Ca^{++}	HC=O	H	GLO	(1)	[1]				
	| HC=O	C, O	GLO	(2/1)	[2/1]	experimental	not	297	64)
		Ca	GLO	(8/4)	[3/2]		given		
Ca^{++}	CHO—CH_2—CHO	H	GLO	(1)	[1]				
		C, O	GLO	(2/1)	[2/1]	assumed	not	389	65)
		Ca	GLO	(8/4)	[3/2]		given		
Ca^{++}	CH_3—CO—NH—CH_3	H	GTO	(3)	[1]				
		C, N, O	GTO	(7/3)	[2/1]	assumed	217	283	98)
		Ca	GTO	(9/6)	[3/2]		Ca—O		

Ca^{++}	$CHO-CH_2-CH_2-CHO$	H C, O Ca	GLO (1) GLO (2/1) GLO (8/4)	[1] [2/1] [3/2]	assumed	not given	464	65)
Ca^{++}	$CH_3-CO-N(CH_3)_2$	H C, N, O Ca	GTO (3) GTO (7/3) GTO (9/6)	[1] [2/1] [3/2]	assumed	218 Ca—O	286	98)
Ca^{++}	$NH_2-CO-CH_2-CO-NH_2$	H C, N, O Ca	GTO (3) GTO (7/3) GTO (9/6)	[1] [2/1] [3/2]	assumed	225 Ca—O	483	98)
Ca^{++}	$NH_2-CO-NH-CO-NH_2$	H C, N, O Ca	GLO (1) GLO (2/1) GLO (8/4)	[1] [2/1] [3/2]	experimental	220 Ca—O	558	71)
Ca^{++}	$CO-N=C(NH_2)-CH=CH-NH$ (ring) cytosine	H C, N, O Ca counterpoise corrected	GLO (1) GLO (2/1) GLO (8/4)	[1] [2/1] [3/2]	experimental	230 Ca—O	439	73)
Ca^{++}	$CH_3-CH_2-CH_2-CO-N(CH_3)_2$	H C, N, O Ca	GTO (3) GTO (7/3) GTO (9/6)	[1] [2/1] [3/2]	assumed	218 Ca—O	294	98)
Ca^{++}	$CO-NH-CH=C(CH_3)-CO-NH$ (ring) thymine	H C, N, O Ca counterpoise corrected	GLO (1) GLO (2/1) GLO (8/4)	[1] [2/1] [3/2]	experimental	217 Ca—O	307	73)
Ca^{++}	$N-C(NH_2)=C-N=CH$ $CH--N==C---NH$ adenine	H C, N Ca counterpoise corrected	GLO (1) GLO (2/1) GLO (8/4)	[1] [2/1] [3/2]	experimental	231 Ca—N	316	73)
Ca^{++}	$NH-CO-C-N=CH$ $NH_2-C==N--C---NH$ guanine	H C, N, O Ca counterpoise corrected	GLO (1) GLO (2/1) GLO (8/4)	[1] [2/1] [3/2]	experimental	233 Ca—N	506	73)
Ca^{++}	$N-CO-CH_2$ $(CH_3)_2N-CO$	H C, N, O Ca	GTO (3) GTO (7/3) GTO (9/6)	[1] [2/1] [3/2]	assumed	224 Ca—O	492	98)

Table 1. (continued)

Ion	Ligand	Basis set[b]				Ligand geometry/ * Complex geometry[c]	R [pm]	—E (Int) [kJ/mole]	Ref.
		atom	type	uncontracted	contracted				
Ca^{++}	$(CH_3)_2N-CO-CH_2$ | $(CH_3)_2N-CO-CH_2$	H C, N, O Ca	GTO GTO GTO	(3) (7/3) (9/6)	[1] [2/1] [3/2]	assumed	224 Ca—O	510	98)

[a] CI and pseudopotential calculations are not included, neither are calculations on outer sphere complexes. Calculations improved by the same authors in subsequent publications using the same basis sets are omitted.

[b] The notation GTO A ($\alpha/\beta/\gamma/\delta$) [a/b/c/d] means: α s-type, β p-type, γ d-type, and δ f-type Gaussian Type Orbitals are centered on atom A which are contracted to a s-type, b p-type, c d-type, and d f-type orbitals respectively.

The notation STO-nG means: a minimal basis set of Slater Type Orbitals is used each of which is expanded into n simple Gaussians [112,113].

The notation n-klG means: each inner shell is represented by a single basis function taken as a sum of n Gaussians while each valence orbital is split into inner and outer parts described by k and l Gaussians, respectively. Asterisks indicate additional polarization functions. For full description see: 3-21G: [114]; 4-31G: [115,116]; 5-21G: [117]; 6-21G: [114]; 6-31G: [117]; 6-31G* and 6-31G**: [118]. For 6G3G see [18].

The counterpoise correction procedure is described in [119]. (See also 2.3.)

[c] The position of the ion was optimized and the geometry of the ligand was not relaxed unless otherwise specified. Entries marked with an asterisk refer to the whole complex, those without to the uncomplexed ligand.

Table 2. Ab initio SCF calculations on some anion-ligand interactions. For details see Table 1.

Ion	Ligand	Basis set				Ligand geometry/ * Complex geometry	R [pm]	−E(Int) [kJ/mole]	Ref.
		atom	type	uncontracted	contracted				
F^-	H_2O	H, O	GTO	STO-4G		STO-4G geometry	240	133	30)
		Li	GTO	?	[2/1]		F—O		
F^-	H_2O	H	GTO	(6/2)	[2/2]				
		O	GTO	(11/7/2)	[4/3/2]	not specified	251	99	33)
		F	GTO	(13/8/2)	[7/4/2]		F—O		
F^-	H_2O	H	GTO	(6/1)	[3/1]	experimental	251	101	35)
		O, F	GTO	(11/7/1)	[5/4/1]		F—O		
F^-	H_2O	H	GTO	(6/1)	[2/1]				
		O	GTO	(11/7/1)	[4/3/1]	not specified	251	98	102)
		F	GTO	(13/8/1)	[7/4/1]		F—O		
F^-	H_2O	all	GTO	4-31G		experimental	248 F—O	168	38)
F^-	$2\ H_2O$	all	GTO	4-31G		* not varied	248 F—O	302	44)
F^-	$2\ H_2O$	H	GTO	(6/1)	[3/1]	experimental	255	188	103)
		O, F	GTO	(11/7/1)	[5/4/1]		F—O		
F^-	$4\ H_2O$	all	GTO	4-31G		* not varied	not given	481	38)
F^-	CH_3—CN	H, C, N	GTO	4-31G		* fully optimized	170	76	104)
		F	GTO	4-31G + p orbital			F—N		
F^-	2 CH_3—CN	H, C, N	GTO	4-31G		4-31G geometry	179	141	104)
		F	GTO	4-31G + p orbital			F—N		
F^-	3 CH_3—CN	H, C, N	GTO	4-31G		4-31G geometry	186	198	104)
		F	GTO	4-31G + p orbital			F—N		
F^-	4 CH_3—CN	H, C, N	GTO	4-31G		4-31G geometry	196	237	104)
		F	GTO	4-31G + p orbital			F—N		
F^-	2 HCOOH	H	GTO	(4/1)	[2/1]	* 4-31G geometry	160	358	105)
		C, O, F	GTO	(9/5)	[4/2]		F—H		

Table 2. (continued)

Ion	Ligand	Basis set				Ligand geometry/ * Complex geometry	R [pm]	−E(Int) [kJ/mole]	Ref.
		atom	type	uncontracted	contracted				
	CO–NH–CH=CH–CO–NH	H	GTO	(4/1)	[2/1]	experimental			
	uracil	C, N, O	GTO	(9/5)	[4/2]	* part. optimized	257	378	106)
		F	GTO	(9/5)	[4/2]		F—N		
Cl^-	H_2O	H	GTO	(6/2)	[2/2]				
		O	GTO	(11/7/1)	[4/3/1]	not specified	331	50	33)
		Cl	GTO	(17/11/2)	[9/6/2]		Cl—O		
Cl^-	H_2O	H	GTO	(6/1)	[2/1]				
		O	GTO	(11/7/2)	[4/3/2]	not specified	331	50	102)
		Cl	GTO	(17/10/1)	[9/6/1]		Cl—O		
Cl^-	H_2O	H, O	GTO	4-31G		fully optimized	240	57	107)
		Cl	GTO	4-31G + p orbital		* part. optimized	Cl—H		
Cl^-	H_2O	all	GTO	4-31G		experimental	315	88	44)
							Cl—O		
Cl^-	H_2O	all	GTO	4-31G		experimental	315	83	38)
							Cl—O		
Cl^-	H_2O	H	GTO	(6/1)	[2/1]				
		O	GTO	(12/8/2)	[5/8/2]	experimental	339	49	78)
		Cl	GTO	(17/11/1)	[10/7/1]		Cl—O		
Cl^-	H_2O	H	STO	(2)					
		O	STO	(3/2)		* part. optimized	304	80	108)
		Cl	STO	(5/4)			Cl—O		
Cl^-	CH_3—OH	H, C, O	GTO	4-31G		fully optimized	237	50	107)
		Cl	GTO	4-31G + p orbital		* part. optimized	Cl—H		
Cl^-	CH_3—CN	H, C, N	GTO	4-31G		* fully optimized	345	48	104)
		Cl	GTO	4-31G + p orbital			Cl—C		
Cl^-	2 CH_3—CN	H, C, N	GTO	4-31G		4-31G geometry	346	91	104)
		Cl	GTO	4-31G + p orbital			Cl—C		
Cl^-	3 CH_3—CN	H, C, N	GTO	4-31G		4-31G geometry	350	128	104)
		Cl	GTO	4-31G + p orbital			Cl—C		

Cl^-	4 CH_3—CN	H, C, N	GTO	4-31G		4-31G geometry	355	158	104)
		Cl	GTO	4-31G + p orbital			Cl—C		
Cl^-	HCOOH	H, C, O	GTO	4-31G		fully optimized	222	92	107)
		Cl	GTO	4-31G + p orbital		* part. optimized	Cl—H		
Cl^-	CH_3—COOH	H, C, O	GTO	4-31G		fully optimized	215	72	107)
		Cl	GTO	4-31G + p orbital		* part. optimized	Cl—H		
Cl^-	$(CH_3)_3C$—OH	H, C, O	GTO	4-31G		fully optimized	244	49	107)
		Cl	GTO	4-31G + p orbital		* part. optimized	Cl—H		
Cl^-	C_6H_5—NH_2	H, C, N	GTO	4-31G		fully optimized	246	47	107)
		Cl	GTO	4-31G + p orbital		* part. optimized	Cl—H		
Cl^-	C_6H_5—OH	H, C, O	GTO	4-31G		fully optimized	227	77	107)
		Cl	GTO	4-31G + p orbital		* part. optimized	Cl—H		
Cl^-	$CHCl_3$	Ligand	GTO	4-31G		fully optimized	230	69	107)
		Cl^-	GTO	4-31G + p orbital		* part. optimized	Cl—H		
Br^-	H_2O	H	GTO	(4)	[2]				
		O	GTO	(9/5)	[4/2]	experimental	339	64	109)
		Br	GTO	(14/11/5)	[8/6/2]		Br—O		
Br^-	H_2O	H	GTO	(4/1)	[2/1]	experimental			
		O	GTO	(9/5/1)	[4/2/1]	* not varied	339	54	109)
		Br	GTO	(14/11/6)	[8/6/3]		Br—O		
Br^-	2 H_2O	H	GTO	(4)	[2]				
		O	GTO	(9/5)	[4/2]	experimental	340	122	109)
		Br	GTO	(14/11/5)	[8/6/2]		Br—O		
Br^-	2 H_2O	H	GTO	(4/1)	[2/1]	experimental			
		O	GTO	(9/5/1)	[4/2/1]	* not varied	340	101	109)
		Br	GTO	(14/11/6)	[8/6/3]		Br—O		
Br^-	3 H_2O	H	GTO	(4)	[2]				
		O	GTO	(9/5)	[4/2]	experimental	345	172	109)
		Br	GTO	(14/11/5)	[8/6/2]		Br—O		
Br^-	4 H_2O	H	GTO	(4)	[2]	experimental			
		O	GTO	(9/5)	[4/2]	* tetrahedral,	344	215	109)
		Br	GTO	(14/11/5)	[8/6/2]	* not optimized	Br—O		
Br^-	4 H_2O	H	GTO	(4)	[2]	experimental			
		O	GTO	(9/5)	[4/2]	* tetragonal	349	213	109)
		Br	GTO	(14/11/5)	[8/6/2]		Br—O		

Table 2. (continued)

Ion	Ligand	Basis set[b] atom	type	uncontracted	contracted	Ligand geometry/ * Complex geometry	R [pm]	−E(Int) [kJ/mole]	Ref.
$HCOO^-$	H_2O	H	GTO	(4/1)	[3/1]	assumed	290	80	110)
		C, O	GTO	(9/5/1)	[5/3/1]		O—O		
$HCOO^-$	H_2O	H	GTO	(3)	[1]	assumed	275	92	110)
		C, O	GTO	(7/3)	[2/1]		O—O		
$HCOO^-$	H_2O	H	GTO	(3)	[1]				
		C, O	GTO	(7/3)	[2/1]	assumed	275	69	110)
		counterpoise corrected					O—O		
HCO_3^-	H_2O	all	GTO	STO-3G		free H_2O and ion	250	103	111)
						fully optimized	O—O		
HCO_3^-	H_2O	all	GTO	4-31G		free H_2O and ion	283	86	111)
						fully optimized	O—O		
HCO_3^-	H_2O	H	GTO	STO-3G		free H_2O and ion	300	95	111)
		C, O	GTO	STO-3G + 3sp orbital		STO-3G geometry	O—O		
HCO_3^-	H_2O	H	GTO	4-31G		free H_2O and ion			
		C, O	GTO	4-31G + 3sp orbital		4-31G geometry	300	79	111)
						* not varied	O—O		
HCO_3^-	2 H_2O	all	GTO	4-31G		free H_2O and ion	285		
						fully optimized	273	161	111)
							O—O		
HCO_3^-	3 H_2O	all	GTO	4-31G		free H_2O and ion	289		
						fully optimized	278	220	111)
							276		
							O—O		

of the host and guest improve each other mutually. Thus $E_{Complex}$ [Eq. (1)] is calculated with a virtually larger basis than E_{Host} and E_{Guest}. Larger basis sets lead to more negative total energies. $E_{Complex}$ is therefore too negative and the interaction energy will be overestimated if calculated according to Eq. (1). This effect was carefully analyzed [123, 125, 126]. Well-balanced small basis sets which are insensitive to the basis set superposition error are now available [129, 130].

The basis set superposition error can be estimated using the counterpoise technique [119]. In this method the energies E_{Host} and E_{Guest} are calculated using the geometry and basis set applied for the complex but with zero charges on the nuclei of the guest and of the host, respectively. It is to be noted that the counterpoise calculation does not exactly eliminate the basis set superposition error. It overestimates the correction term and should not be applied in calculations using extended basis sets [145].

The importance of polarization functions (like d-orbitals for C, O and N or p-orbitals for H) for the calculation of ion-ligand interaction energies was pointed out by several authors [7, 17, 19]. Nevertheles many results using well-balanced small basis sets without polarization functions are acceptable if interactions of ligands with alkali or alkaline earth metal ions are considered (see Tables 1 and 3). The importance of the polarization functions in the case of the ammonium-water interaction was demonstrated recently [91].

2.4 Geometries

For an accurate description of the interaction energy the geometries of the host, the guest and of the complex should be optimized using large basis sets. In practice the time for the computation is prohibitively large except for small systems. A way to circumvent this problem is the use of experimental geometries of the host and guest and to search only for their optimal position and orientation in the complex. In the case of ion-ligand interactions as compiled in Tables 1 and 2 this procedure leads to satisfactory results.

Geometry optimizations using small basis sets are problematic. The calculated geometries may strongly deviate from the experimentel ones. In a recent study of the $H_2O—NH_4^+$ complex, it was shown that the use of experimental geometries may be very satisfactory. The influence of geometry optimization on the calculated interaction energy is in this case insignificant if large basis sets are used, [91] but of significance in case of smaller basis sets [88] (see also Tables 1 and 3).

In case of alkali and alkaline earth metal cations as guests, the calculated ion-ligand distances at the energy minimum depend on the basis set. An improvement of the basis set leads to an increase of the ion-ligand distance. For $Li^+—H_2O$ and $K^+—H_2O$ the optimal distances are 176 and 257 pm, if minimal Gaussian basis sets are used. The corresponding values are 189 and 269 pm with extended basis sets (see Table 1). Even these latter values are significantly lower than the sum of the ionic radii [132] and the van der Waals radius of the oxygen atom (218 pm for Li^+ and 273 pm for K^+). These remaining differences can be explained as a consequence of ligand-ligand repulsions in case of higher coordination number. Ionic radii and van der Waals radii allow, in general, a good approximation of the experimental results if the ions are fully coordinated. They overestimate, however, the optimal distance

Table 3. Comparison of the experimental enthalpies of interaction with the calculated interaction energies as given in Tables 1 and 2.

Ion	Ligand	−ΔH(Int) (exp.) [kJ/mole]	−E(Int) (SCF)[a] [kJ/mole]	Ref. (exp.)	Ref. (SCF)
Li^+	NH_3	164	169	134)	24)
		162		135)	
Li^+	2 NH_3	301	395	135)	27)
Li^+	4 NH_3	458	515	135)	28)
Li^+	5 NH_3	504	723	135)	29)
Li^+	6 NH_3	543	741	135)	29)
Li^+	H_2O	142	147	15)	32)
Li^+	2 H_2O	250	283	15)	43)
Li^+	4 H_2O	405	494	15)	28)
Li^+	5 H_2O	464	671	15)	29)
Li^+	6 H_2O	514	680	15)	29)
Li^+	HCN	152	155	134)	7)
Li^+	H_2CO	151	161	134)	40)
Li^+	CH_3—NH_2	172	167	134)	21)
Li^+	CH_3—OH	159	172	134)	19)
Li^+	CH_3—F	128	146	136)	45)
Li^+	CH_3—CN	180	188	136)	19)
Li^+	CH_3—CHO	174	190	136)	19)
Li^+	CH_3—NH—CH_3	177	182	134)	19)
Li^+	CH_3—O—CH_3	165	172	134)	19)
Li^+	CH_3—Cl	105	87	136)	19)
Li^+	CH_3—CO—CH_3	186	222	136)	52)
Li^+	$N(CH_3)_3$	176	201	134)	18)
Li^+	CH_3—S—CH_3	134	119	136)	19)
Li^+	HCO—$N(CH_3)_2$	209	223	136)	22)
Li^+	CH_3—COO—CH_3	184	172	136)	22)
Li^+	pyridine	185	179	136)	22)
Na^+	NH_3	122	116	135)	20)
Na^+	4 NH_3	351	509	135)	29)
Na^+	5 NH_3	395	585	135)	29)
Na^+	6 NH_3	436	628	135)	29)
Na^+	H_2O	100	105	15)	34)
Na^+	4 H_2O	307	477	15)	29)
Na^+	5 H_2O	359	552	15)	29)
Na^+	6 H_2O	403	594	15)	29)
K^+	NH_3	75	77	137)	20)
		84		138)	
K^+	4 NH_3	257	333	138)	29)
K^+	H_2O	75	70	15)	33)
		71	73	139)	76)
K^+	4 H_2O	247	320	15)	29)
K^+	5 H_2O	292	387	15)	29)
K^+	6 H_2O	333	441	15)	29)
K^+	CH_3—NH_2	80	76	137)	21)
K^+	CH_3—CN	102	107	139)	82)
K^+	CH_3—NH—CH_3	82	79	137)	82)
K^+	CH_3—O—CH_3	87	97	137)	82)
		93		140)	
NH_4^+	NH_3	104	118	141)	85)
		113		142)	

Table 3. (continued)

Ion	Ligand	$-\Delta H(Int)$ (exp.) [kJ/mole]	$-E(Int)$ (SCF)[a] [kJ/mole]	Ref. (exp.)	Ref. (SCF)
NH_4^+	2 NH_3	177	310	141)	84)
		184		142)	
NH_4^+	3 NH_3	235	407	141)	84)
		253		142)	
NH_4^+	4 NH_3	287	475	141)	83)
		314	481	142)	84)
NH_4^+	5 NH_3	318	524	141)	84)
		345		142)	
NH_4^+	H_2O	72	78	141)	91)
NH_4^+	2 H_2O	134	166	141)	88)
NH_4^+	3 H_2O	190	217	141)	88)
NH_4^+	4 H_2O	241	276	141)	88)
NH_4^+	5 H_2O	282	493	141)	84)
F^-	H_2O	97	99	143)	33)
F^-	2 H_2O	167	302	143)	44)
F^-	4 H_2O	281	481	143)	38)
F^-	CH_3-CN	67	76	144)	104)
F^-	2 CH_3-CN	121	141	144)	104)
F^-	3 CH_3-CN	170	198	144)	104)
F^-	4 CH_3-CN	213	237	144)	104)
Cl^-	H_2O	55	50	143)	33)
		62		145)	
Cl^-	CH_3-OH	59	50	146)	107)
Cl^-	CH_3-CN	56	48	144)	104)
Cl^-	2 CH_3-CN	107	91	144)	104)
Cl^-	3 CH_3-CN	151	128	144)	104)
Cl^-	4 CH_3-CN	177	158	144)	104)
Cl^-	HCOOH	155	92	146)	107)
Cl^-	CH_3-COOH	90	72	146)	107)
Cl^-	$(CH_3)_3C-OH$	59	49	146)	107)
Cl^-	$C_6H_5-NH_2$	72	47	143)	107)
Cl^-	C_6H_5-OH	81	77	146)	107)
Cl^-	$CHCl_3$	64	69	146)	107)
Br^-	H_2O	53	54	143)	109)
Br^-	2 H_2O	104	101	143)	109)
Br^-	3 H_2O	152	172	143)	109)
Br^-	4 H_2O	198	215	143)	109)
HCO_3^-	H_2O	66	79	147)	111)
HCO_3^-	2 H_2O	128	161	147)	111)
HCO_3^-	3 H_2O	185	220	147)	111)

[a] Calculated values obtained with the largest basis sets applied were selected. For details and for other calculations see Tables 1 and 2.

in case of a 1:1 complex with a monodentate ligand. The differences are especially large for small cations such as Li^+ and Mg^{2+} [98]. In line with these considerations, a decrease of the metal-oxygen distances with decreasing coordination number was documented recently by comparing a large number of X-ray structures of alkali and alkaline earth metal complexes [133].

Table 4. Examples for the comparison of calculated interaction energies with measured enthalpies (all values in kJ/mol)

Reaction	Temperature [K]	ΔE_e		ΔE_v^o	$\Delta(\Delta E_v)^T$	ΔE_r^T	ΔE_t^T	ΔPV	ΔH_{Calc}^T	ΔH_{Exp}^T	Ref.
		ΔE_{SCF}	ΔE_{CORR}								
H_2O ... H_2O	298	− 18.0	−4.6	9.2	7.9	−3.8	−3.8	−2.5	− 15.5		[148]
	373	− 18.0	−4.6	9.2	9.2	−4.6	−4.6	−2.9	− 16.3	− 15.1 ± 2.1	[148]
H_2O ... Li^+	298	−149.4	2.1	8.8	1.7	0	−3.8	−2.5	−143.1	−142.3 ± 8.4	[148]
	0	−147.3	−1.1	5.6	0	0	0	0	−142.8		[33]
H_2O ... NH_4^+	500	− 78.1	−6.6	13.0	14.3	−6.2	−6.2	−4.2	− 74.0	− 72.4 ± 1.7	[90, 91]

2.5 Correlation Energy

A part of the electronic energy is not considered in case of *ab initio* SCF calculations since the electrons of different spins are treated as independent (uncorrelated) within the framework of this approach. If the corresponding energy (correlation energy) is of different magnitude in the complex and in its constituents, the correlation energy contribution to the interaction energy has to be evaluated.

The intermolecular electron correlation (the dispersion interaction) was calculated or estimated for some cation-ligand interactions using configuration interaction (CI) calculations, perturbation theory or on the basis of a statistical model (see Table 4). Its contribution to the total interaction energy is less than 10% throughout.

2.6 Comparison with the Experiment

Intermolecular interactions in the gas phase have been measured in a series of cases using mass spectrometry [134-147]. From the temperature dependence of the equilibrium constants, besides the free energies, the enthalpies and the entropies of the involved reactions were evaluated. The corresponding data are useful for comparison with the results of theoretical calculations (see Table 3). In order to compare the calculated interaction energies with the measured reaction enthalpies, a series of contributions has to be taken into account. Concerning these correction terms some inconsistencies arise in the literature. Therefore the list of them is given here in detail according to Ref. [148]:

$$\Delta H^{298}_{Calc} = \Delta E^{298}_{Calc} + \Delta PV \tag{4}$$

$$\Delta E^{298}_{Calc} = \Delta E^0_e + \Delta(\Delta E_e)^{298} + \Delta E^0_v + \Delta(\Delta E_v)^{298} + \Delta E^{298}_r + \Delta E^{298}_t \tag{5}$$

ΔE^0_e is the calculated electronic interaction energy at 0 K.
$\Delta(\Delta E_e)^{298}$ is the change in the calculated electronic interaction energy between 298 K and 0 K. This term is negligible unless electronically excited states are important.
ΔE^0_v is the difference between the zero-point vibrational energies of reactants and product.
$\Delta(\Delta E_v)^{298}$ is the change in the vibrational energy difference between 298 K and 0 K.
ΔE^{298}_r is the difference in rotational energies of reactants and product.
ΔE^{298}_t is the translational energy change due to the change in the number of degrees of translational freedom.
The calculation of the vibrational terms is straightforward but rather time consuming (see e.g. [149]).
ΔE^{298}_r and ΔE^{298}_t can be approximated according to classical considerations as $-1/2 \cdot RT$ for each degree of rotational and translational freedom lost due to complex formation. Assuming an ideal behavior, ΔPV is equal to $-RT$ for a 1:1 host-guest association since 1 mol gas is lost by the complexation reaction. In Table 4 the calculated contributions for two simple 1:1 complexes are given.

3 Treatment of Large Systems

3.1 Introduction

The computer time required for *ab initio* calculations is roughly proportional to the fourth power of the number of atomic basis functions used for the description of the molecular system. *Ab initio* calculations are thus not feasible today for host-guest systems with more than about 150—200 electrons. Supercomputers and vector processors will significantly lower the necessary CPU times [150] but they alone probably cannot bring a breakthrough for systems larger than two or three times the ones which can be treated today.

An increase in speed can be achieved by using pseudopotential calculations [122,151]. In these type of calculations the inner shell electrons are approximated with a potential and the problem is reduced to a valence electron problem. This technique is very powerful for heavy atoms but the time saving is not more than roughly 50% for molecules containing first-row atoms only [152]. Ion-ligand interactions have been studied with pseudopotential calculations in several cases [153-157].

In some cases, strange approximations were applied in order to circumvent the problems connected with large systems. The interaction energy of the antibiotic tetranactin with an ammonium ion was calculated by replacing the tetranactin by four formaldehyde and four water molecules [158]. In an "improved" study the tetranactin was approximated by using formic acid, ethane, propane and methanol molecules [159]. In an other study [18] crown-6 was simulated by three dimethyl ether molecules [160].

The most promising methods for the prediciton of interaction energies in realistic host-guest systems which are known at present apply some kind of extrapolation techniques. Results of *ab initio* calculations on small systems are used in these models for the description of large systems. The two approaches which seem to be the most important at present are described in the succeeding chapters.

3.2 Pair Potential Procedure

This technique has widely been applied in a series of papers by Clementi and coworkers for the description of the solvation of amino acids, peptides as well as of RNA, DNA and their constituents (for reviews see [161,162]). The interactions of some cations with these types of molecules were also described [163-166]. Pair potentials between small model molecules and the cations Li^+ [62], Na^+ [167], K^+ [168], NH_4^+ [168], Mg^{2+} [98] and Ca^{2+} [98] were developed in order to describe ion-ionophore interactions [169].

Within the frame of this approach the interaction energy of two molecules is described as the sum of pairwise interactions. It is assumed that each atom of the host interacts with each atom of the guest independently:

$$E_{Int} = \sum_i \sum_j e_{ij} \qquad (5)$$

(i and j are the running numbers of the host and the guest, respectively). This is of course a crude approximation but the reliability of the results confirms the usefulness of this approach. It is very versatile for various reasons. First the calculations are very fast because the only geometrical parameter is the distance r_{ij} between the two atoms, i.e. the pair potentials have spherical symmetry. Angle dependencies are introduced implicitly because the superposition of several pair potentials gives rise to a reduction of the symmetry. A further advantage is that it would be easy to include these simple pair potentials in any molecular mechanics program.

Different forms of the pair potentials e_{ij} can be used. In most cases a three parameter function of the following type was applied:

$$e_{ij} = -A_{ij}/r_{ij}^6 + B_{ij}/r_{ij}^{12} + q_i q_j C_{ij}/r_{ij} \quad (6)$$

A_{ij}, B_{ij} and C_{ij} are constants for a given pair of atoms and q_i and q_j are atomic net charges of the atoms i and j. The first two terms correspond to the Lennard-Jones potential and the third term to the electrostatic point charge — point charge interaction.

In order to evaluate the values of the constants A_{ij}, B_{ij}, and C_{ij}, a large number of *ab initio* calculations are to be made for model systems. In each model system the distances and relative orientations of the constituents are to be varied. The parameters can then be fitted using the Eqs. (5) and (6). Atomic net charges of the uncomplexed host and guest are usually used for this procedure. In general they are calculated with the same basis sets as the interaction energies. Since calculated atomic net charges heavily depend on the basis sets (small basis sets tend to overestimate the polarization), a parameter set can only be applied by using net charges obtained with the same basis set as for the fitting procedure.

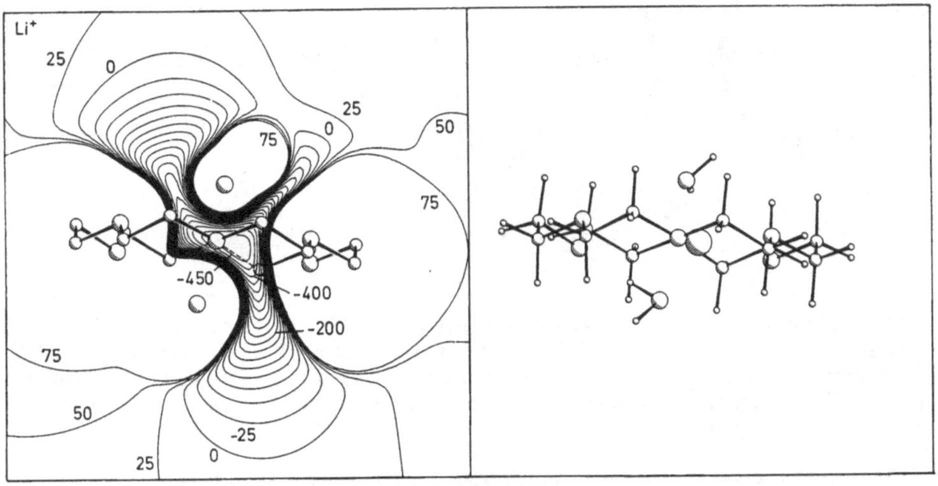

Fig. 1. Left: Isoenergy countour diagram (energies in kJ/mol) for the interaction of Li^+ with [18] crown-6 and 2 H_2O [75]. The conformation of the crown ether and the position of the water molecules were fixed as found experimentally [170]. Right: Same view of the structure of the $LiClO_4 \cdot$ [18] crown-6 $\cdot$ 2 H_2O complex as determined by X-ray crystallography [170]. The distance between the position of the energy minimum (left) and the found position of Li^+ (right) amounts to 7 pm

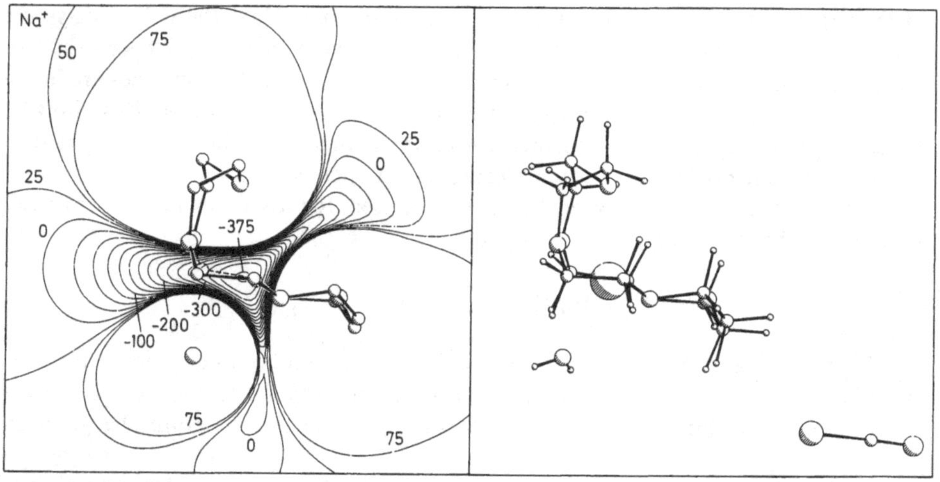

Fig. 2. Left: Isoenergy contour diagram (energies in kJ/mol) for the interaction of Na^+ with [18] crown-6 and one H_2O molecule [75]. The conformation of the crown ether and the position of the water molecule were fixed as found experimentally [171]. Right: same view of the structure of the NaSCN · [18] crown-6 · H_2O complex as determined by X-ray crystallography [171]. The distance between the position of the energy minimum (left) and the found position of Na^+ (right) amounts to 11 pm

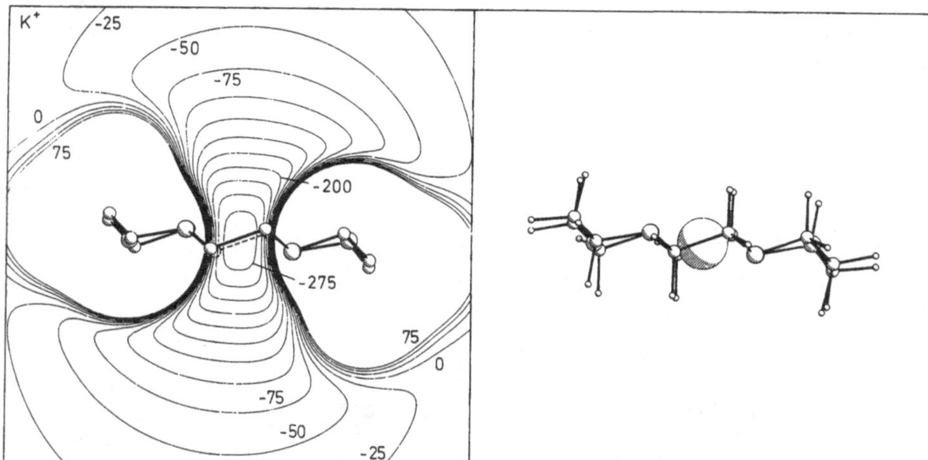

Fig. 3. Left: Isoenergy contour diagram (energies in kJ/mol) for the interaction of K^+ with [18] crwon-6 [75]. The conformation of the crown ether was fixed as found experimentally [172]. Right: Same view of the structure of the KSCN · [18] crown-6 complex as determined by X-ray crystallography [172]. The distance between the position of the energy minimum (left) and the found position of K^+ amounts to 0 pm

Atoms of the same kind in similar chemical environments can be grouped in the same classes, i.e. they are forced to have the same constants. By this procedure a compromise between flexibility and accuracy can be made. Depending on the purpose of the calculations different class assignments and thus different sets of constants can be fitted on the basis of the same set of *ab initio* interaction energies.

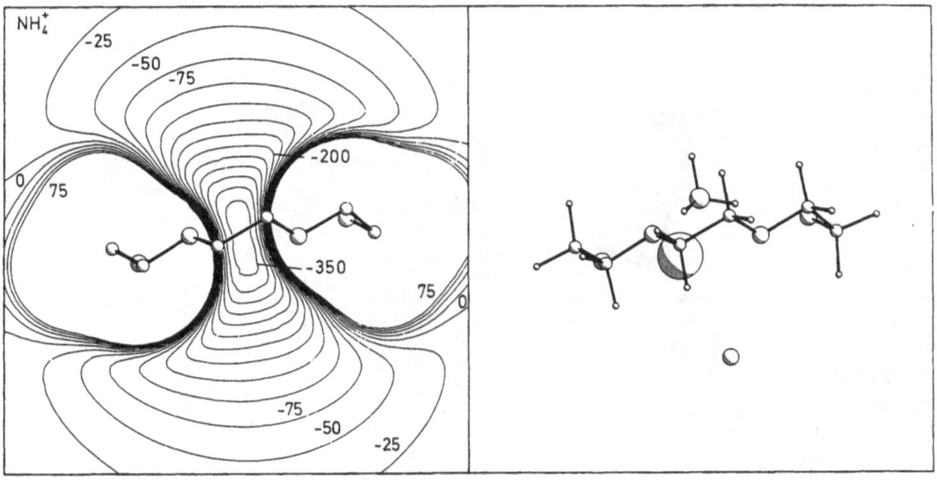

Fig. 4. Left: Isoenergy contour diagram (energies in kJ/mol) for the interaction of NH_4^+ with [18] crown-6 [75]. The conformation of the crown ether was fixed as found experimentally [173]. Right: Same view of the structure of the $NH_4Br \cdot H_2O \cdot$ [18] crown-6 complex as determined by X-ray crystallography [173]. The distance between the position of the energy minimum (left) and the found position of NH_4^+ (right) amounts to 50 pm

In order to describe the interaction energies of large host-guest systems for each atom pair i–j, a corresponding similar atom pair has to be selected from the model systems. Furthermore, the atomic net charges have to be calculated or estimated on the basis of a model compound.

The reliability of the model in predicting the geometry of the complexes is documented in the Figs. 1–5. In these examples the interaction energies of crown ethers with different ions were calculated for 10'000 points of a selected plane. The isoenergy contour diagrams are depicted together with the structures as determined by X-ray crystallography. For the interaction energy calculations the geometry of the ligands was fixed in the conformation as found experimentally. The deviation between experimental and calculated positions of the ions is 0–50 pm.

Similar pair potentials were succesfully used in many applications involving interaction energy calculations on channels [165,166], DNA [163], and also including Monte Carlo techniques [163,165,166,175,176].

3.3 Additive Model of Gresh, Claverie and Pullman

In contrast to the pair potential model where the interaction energy hypersurface is approximated by an additive procedure using a simple mathematical function, the basis of the model of Gresh et. al. is an energy partitioning scheme.

The total interaction energy may be described as a sum of contributions (see e.g. [177]):

$$E_{INT} = E_{COU} + E_{EX} + E_{POL} + E_{DISP} + E_{CT} \tag{7}$$

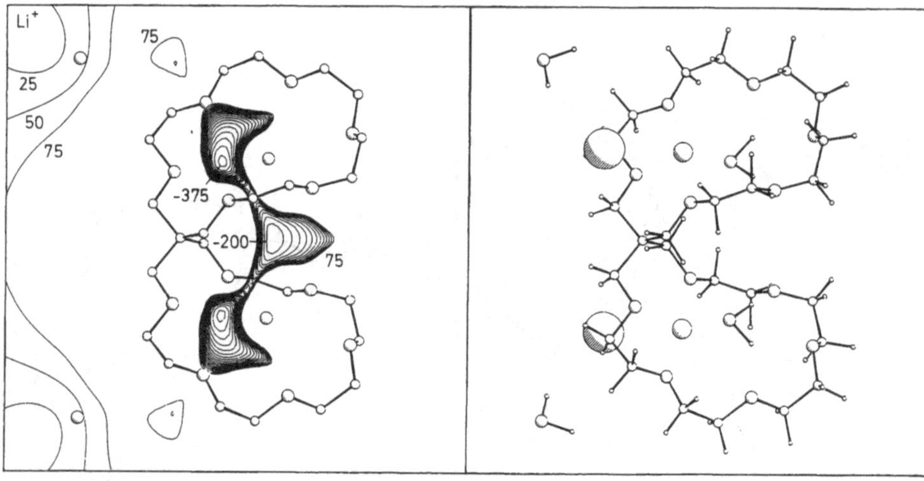

Fig. 5. Left: Isoenergy contour diagram (energies in kJ/mol) for the interaction of Li^+ with 18,18′-spirobi-([19] crown-6) and 4 H_2O [75]. The conformation of the crown ether and the position of the water molecules were fixed as found experimentally [174]. Right: Same view of the structure of the $(LiI)_2$ · 18,18′-spirobi-([19] crown-6) · 4 H_2O complex as determined by X-ray crystallography [174a]. The distance between the position of the energy minimum (left) and the found position of Li^+ (right) amounts to 50 pm. This deviation might be mainly due to the Li^+-Li^+ interaction which was not considered in the calculations

E_{COU} is the electrostatic interaction energy as calculated on the basis of the charge distribution of the isolated host and guest.
E_{EX}, the exchange energy, is a repulsive contribution due to the overlap of the electron densities of host and guest.
E_{POL} is the polarization energy which is a stabilizing term due to the relaxation of the electron density of host and guest in the field of the partner.
E_{DISP} is the dispersion energy which is due to the correlation of electron movements of host and guest.
E_{CT} is the charge transfer energy.

The first four terms result automatically if the interaction energy is calculated by a perturbation treatment (see e.g. [178]). The interaction energy calculated by the *ab initio* SCF technique may be divided into the above contributions (except E_{DISP} which corresponds to the correlation energy) according to model considerations [179,180].

The model proposed by Gresh et.al. [181] approximates the individual terms in the formalism (7) as follows:

$$E_{INT} = E_{MTP} + E_{REP} + E_{POL} + E_{DL} \tag{8}$$

E_{MTP} is the electrostatic interaction energy calculated as a sum of multipole-multipole interactions using the overlap multipole expansion of the SCF electron density distributions of the host and guest [182].
E_{REP} is the sum of bond-bond (or bond-ion) repulsive interactions. $E_{MTP} + E_{REP}$ corresponds to $E_{COU} + E_{EX}$ in Eq. (7).
E_{POL} is the polarization energy.
E_{DL} approximates $E_{DISP} + E_{CT}$ in Eq. (7).

Approximate formulas are used for the calculation of E_{REP}, E_{POL}, and E_{DL}. Parameters used in these approximations are estimated on the basis of *ab initio* calculations on a few small model systems [181]. E_{MTP} brings about the major contribution of the total interaction energy so that uncertainties in the other terms are of only minor importance. For larger molecules E_{MTP} is computed on the basis of *ab initio* calculations on subunits [183].

This additive procedure was applied for the study of a number of cases including the interaction of cations with the carrier antibiotics valinomycin [184] and nonactin [185], the interaction of $CH_3NH_3^+$ and $(CH_3)_4N^+$ with amino acids mimicking the active site of a phosphorylcholine antibody [186], the interaction of guanine and cytosine with amino acids [187], the interaction of Ca^{2+} and Mg^{2+} with two serine phosphates [188], and the interaction of the channel-forming antibiotic gramicidin A with different cations [189–191].

3.4 Comparison of the Two Models

It is not easy to directly compare the two models which were discussed in the previous sections. The only interactions which were studied with both techniques are those of a gramicidin A channel with Na^+ and K^+. In Fig. 6 the results of these studies [165,166,189,191] are compared. The largest differences can be observed at the two ends of the channel. This is due to the fact that the ethanolamine tails were fixed in different con-

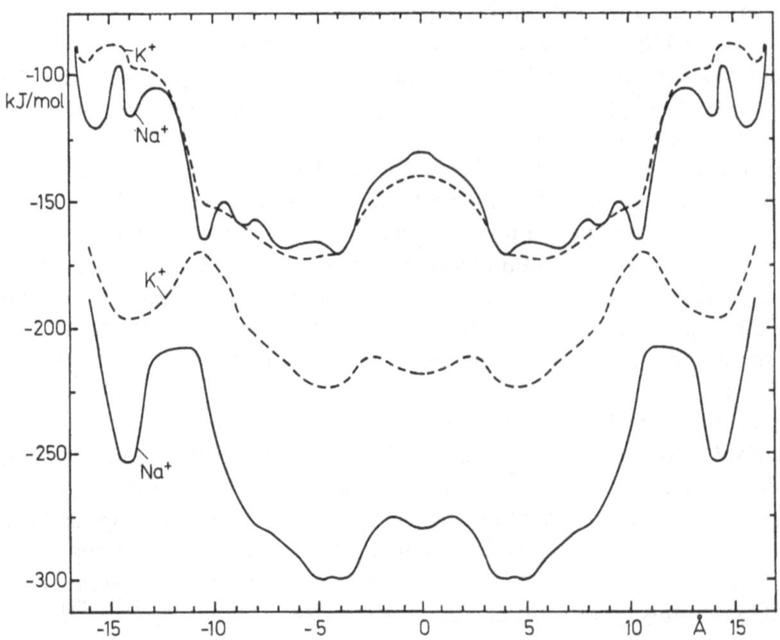

Fig. 6. Minimum interaction energies of K^+ and Na^+ with gramicidin A as a function of their position along the channel axis. The top two curves [191] are calculated according to a model proposed by Gresh et. al. with blocked ethanolamine end chain. The more attractive curves [165,166] are determined using the pair potential method. The gramicidin A dimer ranges from about −14 Å to +14 Å

formations in the two sets of calculations. Similar trends of the two sets of curves are obvious inside the channel. Differences may be due to the fact that the geometries are possibly not exactly the same (Ref. [192] and Ref. [193]) and that in one case [189] the peptide chain was approximated by a polyglycine chain. There is, however, no obvious reason for the drastic differences in the absolute values of the interaction energies (roughly a factor of 2 in case of Na^+). The difference between the interaction energies of the two ions is very pronounced in one case [165, 166] and practically vanishes in the other case [189,191].

The evaluation of pair potentials is much more time-consuming than the evaluation of the parameters in the model of Gresh et al. On the other hand, pair potentials are easy to transfer and corresponding interaction energy calculations are very fast. In contrast, the application of the model of Gresh et. al. includes always time-demanding *ab initio* calculations.

4 Conformational Energy

Although the topic of the present paper is the calculation of interaction energies, we have to treat briefly the contribution of the conformational energy. Isolated hosts and guests exhibit, in general, conformations different from those observed in complexes. A part of the interaction energy is thus needed to bring the host and guest molecules into the appropriate conformation. For the prediction of the overall interaction energy as well as of the structure of the complexes, reliable calculations of the conformational energies are therefore absolutely necessary.

Ab initio calculations using not too small basis sets would be adequate if the computational demands were not prohibitively large. For an accurate geometry optimization the relaxation of all parameters is necessary. Today such calculations are only practicable for relatively small systems.

Very recently the additive procedure of Gresh et al. which was discussed in Sect. 3.3 was extended for the calculation of conformational energy variations in large molecules [194]. The molecules are built up out of constitutive fragments and the intramolecular energy is calculated as a sum of interaction energies between the fragments. The results published so far are very promising. Although the necessary computational demands are substantially lower than for *ab initio* calculations (proportional to n^2 instead of roughly n^4, n being the total number of atoms in the system) they are still significant.

Semiempirical quantum chemical calculations are still too much time-consuming for larger systems. Out of the numerous methods PCILO (Perturbation Configuration Interaction using Localized Orbitals [195]) was proposed by different authors to be the most reliable procedure for conformational analysis (see e.g. [196]). It was applied for many conformational energy studies (for references see e.g. [197]). Comparisons made with *ab initio* and experimental results have however shown in several cases that also PCILO gives only crude estimates of the conformational energy [197–202].

A vast amount of empirical molecular potential energy functions and a series of corresponding programs (molecular mechanics and consistent force field programs) are available (for recent reviews see [203–205]). Unfortunately these energy functions are always the result of optimization on a rather limited group of compounds. No parameter set is available for general use today. An excellent review describing briefly the most important contributions appeared recently [205].

It lies in the nature of the method that most practical applications are extrapolations using parameters optimized on a known set of observables. It is therefore simply not possible to quantitatively predict the reliability of the results. In order to moderate too optimistic expectations we collected some results on [18] crown-6 and its complexes which were obtained with three different molecular mechanics programs (see Table 5). The conformational energies were calculated relative to a minimum energy conformation obtained by relaxation of the experimentally observed structure of the uncomplexed ligand [208]. The reference structures obtained with these methods vary significantly (differences of the corresponding torsion angles of up to 20° were obtained). As shown in Table 5, even the stability sequence of the conformations is inconsistent.

Table 5. Conformation energy calculations on [18] crown-6 with different molecular mechanics methods

Conformation	Conformational energy [kJ/mol][a]		
	WBFF [206]	AMBER [207]	MM2 [75]
C_1 (Na^+-complex)[b]	18.4	39.3	28.0
D_{3d} (K^+-complex)[c]	32.8	4.6	−10.5

[a] Conformational energies are given relative to the energy minimum obtained by relaxation of the experimentally observed structure of the free ligand.
[b] Conformation obtained by the relaxation of the experimentally observed conformation of the ligand in its Na^+-complex.
[c] Conformation obtained by the relaxation of the experimentally observed conformation of the ligand in its K^+-complex.

Such discouraging results should by no means suggest that this type of calculations is of no help for designing hosts. Although they can fail in the quantitative prediction of conformation energies, such calculations can be used to predict in a qualitative way whether a designed molecule has a chance to be a host for a selected guest or not. The usefulness of empirical energy functions for designing macrocyclic ionophores was demonstrated recently by Lifson et al. [6]. Although only estimated parameters of the cations Li^+, Na^+ and K^+ were used, the model was successful in the prediction of ionophoric capability and incapability of different members of the compound class studied [6].

5 Future Prospects

All the calculations of interaction energies of host-guest systems, as discussed above, refer to isolated species in the gas phase. For practical purposes, values in solutions are of interest. Besides interaction energies and conformation energies, the solvation effects of all participants should be included. For ionophores as hosts, the interaction with the counterion of the ionic guest should also be considered. It is clear that already a precise calculation of the two most important terms is problematic. The estimation of the remaining terms, which were not discussed in this paper, is even more difficult. These facts might lead to a pessimistic judgement of the practical value of model calculations.

Far from this pessimism we are convinced that calculations using today's possibilities with all of the limitations are useful as a design aid for hosts. For a ligand design the question should not be "what is the magnitude of the interaciton energy" but rather "has my planned compound a chance to be a host for the selected guest or not". In many cases one tried to answer this question by building molecular models. Today's possibilities of software and hardware for molecular modelling allow a big step forward. The molecules can be built up at the computer terminal and primitive models of conformational constraints and optimization are available. Already such primitive models allow an estimation of the complexing capability of the designed compounds according to the concept put forward by Lifson et. al. [6]. For the design of ionophores as hosts pair potentials developed on the basis of *ab initio* calculations (see Chapter 3.2) could easily be combined with an existing parameter set for conformation energy calculations.

No precise prediction of experimental interaction energies between realistic hosts and guests in solutions is to be expected in the near future. Calculations using existing models can however be used as a design aid and might prevent the synthesis of a large number of planned hosts which are hopeless candidates. It is to be expected that through the availability and the increasing popularity of molecular modelling systems this type of computer aided design will be routinely used within a few years.

6 Acknowledgements

This work was partly supported by the „Schweizerischer Nationalfonds zur Förderung wissenschaftlicher Forschung". We thank PD Dr. T.-K. Ha for critical reading of the manuscript and for valuable comments.

7 References

1. Vögtle, F., Weber, E. (eds.): Host Guest Complex Chemistry I—III, Top. Curr. Chem. *98*, *101*, *121*, Springer-Verlag, Berlin, Heidelberg, New York, 1981, 1982, 1984
2. a) Morf, W. E., Simon, W.: Helv. Chim. Acta *54*, 794 (1971)
   b) Morf, W. E., Simon, W.: ibid *54*, 2683 (1971)
3. Simon, W., Morf, W. E., Meier, P. Ch.: Struct. Bonding (Berlin) *16*, 114 (1973)
4. Pretsch, E., Ammann, D., Simon, W.: Res. Development *25*, 20 (1974)
5. Morf, W. E., Ammann, D., Bissig, R., Pretsch, E., Simon, W., in: Progress in Macrocyclic Chemistry (eds. Izatt, R. M., Christensen, J. J.), Wiley-Interscience, New York 1979
6. Lifson, S., Felder, C. E., Shanzer, A.: J. Am. Chem. Soc. *105*, 3866 (1983)
7. Schuster, P., Jakubetz, W., Marius, W.: Top. Curr. Chem. *60*, 1 (1975)
8. Pullman, B., Goldblum, N. (Eds.): Metal-Ligand Interactions in Organic Chemistry and Biochemistry, D. Reidel Publishing Company, Dordrecht 1977
9. Hobza, P., Zahradnik, R.: Weak Intermolecular Interactions in Chemistry and Biology, Elsevier, Amsterdam 1980

10. Ratajczak, M., Orville-Thomas, W. J. (Eds.): Molecular Interactions, John Wiley, Chichester 1980
11. Scheraga, H. A.: Adv. Phys. Org. Chem. *6*, 103 (1968)
12. Shipman, L. L., Burgess, A. W., Scheraga, H. A.: Proc. Nat. Acad. Sci. *72*, 543 (1975)
13. Nemethy, G., Pottle, M. S., Scheraga, H. A.: J. Phys. Chem. *87*, 1883 (1983)
14. Simon, W., Morf, W. E., Ammann, D., in: Calcium Binding Proteins and Calcium Function, p. 50 (eds. Wasserman, R. M. et al.), Elsevier, Amsterdam 1977
15. Džidić, I., Kebarle, P.: J. Phys. Chem. *74*, 1466 (1970)
16. Scrocco, E., Tomasi, J.: Top. Curr. Chem. *42*, 95 (1973)
17. Woodin, R. L., Houle, F. A., Goddard III, W. A.: Chem. Phys. *14*, 461 (1976)
18. Pullman, A., Brochen, P.: Chem. Phys. Lett. *34*, 7 (1975)
19. Smith, S. F., Chandrasekhar, J., Jorgensen, W. L.: J. Phys. Chem. *86*, 3308 (1982)
20. Berthod, H., Pullman, A.: Chem. Phys. Lett. *70*, 434 (1980)
21. Berthod, H., Pullman, A.: Isr. J. Chem. *19*, 299 (1980)
22. Hinton, J. F., Beeler, A., Harpool, D., Briggs, R. W., Pullman, A.: Chem. Phys. Lett. *47*, 411 (1977)
23. Kollman, P., Rothenberg, S.: J. Am. Chem. Soc. *99*, 1333 (1977)
24. Hinchliffe, A., Dobson, J. C.: Theor. Chim. Acta *39*, 17 (1975)
25. Umeyama, H., Morokuma, K.: J. Am. Chem. Soc. *99*, 1316 (1977)
26. Støgård, Å.: Acta Chem. Scand. *27*, 2669 (1973)
27. Nicely, V. A., Dye, J. L.: J. Chem. Phys. *52*, 4795 (1970)
28. Ribas Prado, F., Giessner-Prettre, C., Daudey, J.-P., Pullman, A., Hinton, J. F., Young, G., Harpool, D.: J. Magn. Res. *37*, 431 (1980)
29. Nagata, C., Aida, M.: J. Theor. Biol. *110*, 569 (1984)
30. Breitschwerdt, K. G., Kistenmacher, H.: Chem. Phys. Lett. *14*, 288 (1972)
31. Pullman, A., Berthod, H., Gresh, N.: Int. J. Quant. Chem. Symp. *10*, 59 (1976)
32. Clementi, E., Popkie, H.: J. Chem. Phys. *57*, 1077 (1972)
33. Kistenmacher, H., Popkie, H., Clementi, E.: ibid *59*, 5842 (1973)
34. Diercksen, G. H. F., Kraemer, W. P.: Theor. Chim. Acta *23*, 387 (1972)
35. Diercksen, G. H. F., Kraemer, W. P., Roos, B. O.: ibid *36*, 249 (1975)
36. Kollman, P. A., Kuntz, I. D.: J. Am. Chem. Soc. *94*, 9236 (1972)
37. Bauge, K., Støgård, Å.: Acta Chem. Scand. *27*, 2683 (1973)
38. Kollman, P., Kuntz, I.: J. Am. Chem. Soc. *98*, 6820 (1976)
39. Schuster, P., Preuss, H.-W.: Chem. Phys. Lett. *11*, 35 (1971)
40. Schuster, P., Marius, W., Pullman, A., Berthod, H.: Theor. Chim. Acta *40*, 323 (1975)
41. Rode, B. M., Sagarik, K. P.: Chem. Phys. Lett. *88*, 337 (1982)
42. Raghavachari, K.: J. Chem. Phys. *76*, 5421 (1982)
43. Kraemer, W. P., Diercksen, G. H. F.: Theor. Chim. Acta *23*, 393 (1972)
44. Kollman, P.: J. Am. Chem. Soc. *99*, 4875 (1977)
45. Del Bene, J. E., Frisch, M. J., Raghavachari, K., Pople, J. A., v. R. Schleyer, P.: J. Phys. Chem. *87*, 73 (1983)
46. Russegger, P., Schuster, P.: Chem. Phys. Lett. *19*, 254 (1973)
47. Neumann, D. B., Moskowitz, J. W.: J. Chem. Phys. *50*, 2216 (1969)
48. Raber, D. J., Raber, N. K., Chandrasekhar, J., von Ragué Schleyer, P.: Inorg. Chem. *23*, 4076 (1984)
49. Ha, T.-K., Wild, U. P., Kühne, R. O., Loesch, Ch., Schaffhauser, T., Stachel, J., Wokaun, A.: Helv. Chim. Acta *61*, 1193 (1978)
50. Schuster, P., Pullman, A.: Chem. Phys. Lett. *24*, 472 (1974)
51. Armbruster, A. M., Pullman, A.: FEBS Letters *49*, 18 (1974)
52. Del Bene, J. E.: Chem. Phys. *40*, 329 (1979)
53. Del Bene, J. E.: Chem. Phys. Lett. *64*, 227 (1979)
54. Nowek, A., Leszczyński, J., Wojciechowski, W.: Materials Science *9*, 71 (1983)
55. Douglas, J., Kollman, P.: J. Phys. Chem. *85*, 2718 (1981)
56. Sapse, A. M., Bunce, J. D., Jain, D. C.: J. Am. Chem. Soc. *106*, 6579 (1984)
57. Rode, B. M., Preuss, H.: Theor. Chim. Acta *35*, 369 (1974)
58. Rode, B. M., Breuss, M., Schuster, P.: Chem. Phys. Lett. *32*, 34 (1975)
59. Sagarik, K. P., Rode, B. M.: Z. Naturforsch. *36a* 1357 (1981)

60. Rode, B. M.: Chem. Phys. Lett. *26*, 350 (1974)
61. Pullman, A., Giessner-Prettre, C., Kruglyak, Yu. V.: ibid *35*, 156 (1975)
62. Corongiu, G., Clementi, E., Pretsch, E., Simon, W.: J. Chem. Phys. *72*, 3096 (1980)
63. Hertz, H. G., Weingärtner, H., Rode, B. M.: Ber. Bunsenges. Phys. Chem. *79*, 1190 (1975)
64. Rode, B. M., Kraft, H. G.: Chem. Phys. Lett. *61*, 410 (1979)
65. Kraft, H. G., Rode, B. M.: Monatsh. Chem. *111*, 797 (1980)
66. Blomberg, M. R. A., Fischer-Hjalmars, I., Henriksson-Enflo, A.: Isr. J. Chem. *19*, 143 (1980)
67. Fuchs, D. N., Rode, B. M.: Chem. Phys. Lett. *82*, 517 (1981)
68. Sagarik, K. P., Rode, B. M.: Z. Naturforsch. *39a*, 686 (1984)
69. Rode, B. M., Fussenegger, R.: J. Chem. Soc., Faraday Trans. 2 *71*, 1958 (1975)
70. Pretsch, E., Neszmelyi, A., Simon, W., Corongiu, G., Clementi, E.: IBM Research Report POK-07; available from IBM, Department B28, Building 701, PO Box 390, Poughkeepsie, New York 12602
71. Rode, B. M., Gstrein, K. H.: J. Chem. Soc., Faraday Trans. 2 *74*, 889 (1978)
72. Sagarik, K. P., Rode, B. M.: Inorg. Chim. Acta *78*, 81 (1983)
73. Sagarik, K. P., Rode, B. M.: ibid *76*, L209 (1983)
74. Rode, B. M., Hannongbua, S. V.: ibid *96*, 91 (1985)
75. Welti, M., Portmann, P., Badertscher, M. Neszmelyi, A., Clementi, E., Pretsch, E., Simon, W.: in preparation
76. Kistenmacher, H., Popkie, H., Clementi, E.: J. Chem. Phys. *58*, 1689 (1973)
77. Perricaudet, M., Pullman, A.: FEBS Letters *34*, 222 (1973)
78. Dacre, P. D.: Mol. Phys. *51*, 633 (1984)
79. Fussenegger, R., Rode, B. M.: Chem. Phys. Lett. *44*, 95 (1976)
80. Perahia, D., Pullman, A., Pullman, B.: Theor. Chim. Acta *43*, 207 (1977)
81. Berthod, H., Pullman, A.: ibid *47*, 59 (1978)
82. Kollman, P.: Chem. Phys. Lett. *55*, 555 (1978)
83. Pullman, A., Armbruster, A. M.: Int. J. Quant. Chem. Symp. *8*, 169 (1974)
84. Pullman, A., Armbruster, A. M.: Chem. Phys. Lett. *36*, 558 (1975)
85. Delpuech, J.-J., Serratrice, G., Strich, A., Veillard, A.: Mol. Phys. *29*, 849 (1975)
86. Merlet, P., Peyerimhoff, S. D., Buenker, R. J.: J. Am. Chem. Soc. *94*, 8301 (1972)
87. Baird, N. C.: Int. J. Quant. Chem.: Quant. Biol. Symp. *1*, 49 (1974)
88. Ikuta, S.: Chem. Phys. Lett *95*, 604 (1983)
89. Böhm, H.-J., McDonald, I. R.: J. Chem. Soc., Faraday Trans. 2 *80*, 887 (1984)
90. Pullman, A., Claverie, P., Cluzan, M.-C.: Chem. Phys. Lett. *117*, 419 (1985)
91. Welti, M., Ha, T.-K., Pretsch, E.: J. Chem. Phys. *83*, 2959 (1985)
92. Timko, J. M., Moore, S. S., Walba, D. M., Hiberty, P. C., Cram, D. J.: J. Am. Chem. Soc. *99*, 4207 (1977)
93. Umeyama, H., Nomoto, T.: Chem. Pharm. Bull. *27*, 1112 (1979)
94. Sano, M., Yamatera, H., in: Ions and Molecules in Solution (eds. Tanaka, N., Ohtaki, H., Tamamushi, R.), Studies in Physical and Theoretical Chemistry *27*, 109 (1982)
95. Corongiu, G., Clementi, E.: J. Chem. Phys. *69*, 4885 (1978)
96. Kochanski, E., Prissette, J.: Chem. Phys. Lett. *80*, 564 (1981)
97. Demoulin, D., Fischer-Hjalmars, I., Henriksson-Enflo, A.: Int. J. Quant. Chem. *12* Suppl. 1, 351 (1977)
98. Welti, M., Pretsch, E., Clementi, E., Simon, W.: Helv. Chim. Acta *65*, 1996 (1982)
99. Perahia, D., Pullman, A., Pullman, B.: Theor. Chim. Acta *42*, 23 (1976)
100. Lavery, R., Pullman, B.: Int. J. Quant. Chem.: Quant. Biol. Symp. *6*, 467 (1979)
101. Gottschalk, K. E., Hiskey, R. G., Pedersen, L. G., Koehler, K. A.: J. Mol. Struct. THEOCHEM *90*, 265 (1982)
102. Kistenmacher, H., Popkie, H., Clementi, E.: J. Chem. Phys. *58*, 5627 (1973)
103. Kraemer, W. P., Diercksen, G. H. F.: Theor. Chim. Acta *27*, 265 (1972)
104. Yamabe, S., Hirao, K.: Chem. Phys. Lett *84*, 598 (1981)
105. Emsley, J., Jones, D. J., Osborn, R. S., Overill, R. E.: J. Chem. Soc., Dalton Trans. 809 (1982)
106. Emsley, J., Jones, D. J., Overill, R. E.: J. Chem. Soc., Chem. Commun. 476 (1982)
107. Yamabe, S., Ihira, N., Hirao, K.: Chem. Phys. Lett. *92*, 172 (1982)
108. Piela, L.: ibid *19*, 134 (1973)
109. Ikuta, S.: ibid *68*, 179 (1979)

110. Berthod, H., Pullman, A.: J. Comput. Chem. *2*, 87 (1981)
111. Jean, Y., Volatron, F.: Chem. Phys. *53*, 95 (1980)
112. Hehre, J. W., Stewart, R. F., Pople, J. A.: J. Chem. Phys. *51*, 2657 (1969)
113. Hehre, J. W., Ditchfield, R., Stewart, R. F., Pople, J. A.: ibid *52*, 2769 (1970)
114. Binkley, J. S., Pople, J. A., Hehre, W. J.: J. Am. Chem. Soc. *102*, 939 (1980)
115. Ditchfield, R., Hehre, W. J., Pople, J. A.: J. Chem. Phys. *54*, 724 (1971)
116. Hehre, W. J., Lathan, W. A.: ibid *56*, 5255 (1972)
117. Dill, J. D., Pople, J. A.: ibid *62*, 2921 (1975)
118. Hariharan, P. C., Pople, J. A.: Theor. Chim. Acta *28*, 213 (1973)
119. Boys, S. F., Bernardi, F.: Mol. Phys. *19*, 553 (1970)
120. Daudey, J. P., Novaro, O., Kołos, W., Berrondo, M.: J. Chem. Phys. *71*, 4297 (1979)
121. Clementi, E., Kistenmacher, H., Kołos, W., Romano, S.: Theor. Chim. Acta *55*, 257 (1980)
122. Carsky, P., Urban, M.: Lecture Notes in Chemistry, Vol. 16, Springer-Verlag, Berlin 1980
123. Pullman, B., Pullman, A., Berthod, H., Gresh, N.: Theor. Chim. Acta *40*, 93 (1975)
124. Alagona, G., Ghio, C., Kollman, P.: J. Am. Chem. Soc. *105*, 5226 (1983)
125. Kołos, W.: Theor. Chim. Acta *51*, 219 (1979)
126. Kołos, W.: ibid *54*, 187 (1980)
127. Poirier, R., Kari, R., Csizmadia, I. G.: Handbook of Gaussian Basis Sets, Elsevier, Amsterdam, 1985
128. Clementi, E.: J. Chem. Phys. *46*, 3851 (1967)
129. Gianolio, L., Pavani, R., Clementi, E.: Gazz. Chim. Ital. *108*, 181 (1978)
130. Gianolio, L., Clementi, E.: ibid *110*, 179 (1980)
131. Carravetta, V., Clementi, E.: J. Chem. Phys. *81*, 2646 (1984)
132. Goldschmidt, V. M.: Skrifter Norske Videnskaps-Akad. Oslo, I., Mat.-Naturuis. Kl. (1926)
133. Chakrabarti, P., Venkatesan, K., Rao, C. N. R.: Proc. Royal Soc. (London) *A375*, 127 (1981)
134. Woodin, R. L., Beauchamp, J. L.: J. Am. Chem. Soc. *100*, 501 (1978)
135. Castleman Jr., A. W., Holland, P. M., Lindsay, D. M., Peterson, K. I.: ibid *100*, 6039 (1978)
136. Staley, R. H., Beauchamp, J. L.: ibid *97*, 5920 (1975)
137. Davidson, W. R., Kebarle, P.: ibid *98*, 6133 (1976)
138. Castleman Jr., A. W.: Chem. Phys. Lett. *53*, 560 (1978)
139. Davidson, W. R., Kebarle, P.: J. Am. Chem. Soc. *98*, 6125 (1976)
140. Davidson, W. R., Kebarle, P.: Can. J. Chem. *54*, 2594 (1976)
141. Payzant, J. D., Cunningham, A. J., Kebarle, P.: ibid *51*, 3242 (1973)
142. Searles, S. K., Kebarle, P.: J. Phys. Chem. *72*, 742 (1968)
143. Arshadi, M., Yamdagni, R., Kebarle, P.: ibid *74*, 1475 (1970)
144. Yamdagni, R., Kebarle, P.: J. Am. Chem. Soc. *94*, 2940 (1972)
145. Castleman Jr., A. W., Holland, P. M., Keesee, R. G.: Radiat. Phys. Chem. *20*, 57 (1982)
146. Yamdagni, R., Kebarle, P.: J. Am. Chem. Soc. *93*, 7139 (1971)
147. Keesee, R. G., Lee, N., Castleman Jr., A. W.: ibid *101*, 2599 (1979)
148. Del Bene, J. E., Mettee, H. D., Frisch, M. J., Luke, B. T., Pople, J. A.: J. Phys. Chem. *87*, 3279 (1983)
149. Pople, J. A., Schlegl, H. B., Krishnan, R., DeFrees, D. J., Binkley, J. S., Frisch, M. J., Whiteside, R. A., Hout, R. J., Hehre, W. J.: Int. J. Quant. Chem. Symp. *13*, 325 (1979)
150. Clementi, E., Corongiu, G., Detrich, J., Chin, S., Domingo, L.: ibid *18*, 601 (1984)
151. Melius, C. F., Goddard III, W. A.: Phys. Rev. *A10*, 1528 (1974)
152. Topiol, S., Ratner, M. A., Moskowitz, J. W.: Chem. Phys. *20*, 1 (1977)
153. Marius, W., Schuster, P.: Theor. Chim. Acta *42*, 5 (1976)
154. Pullman, A., Gresh, N., Daudey, J. P., Moskowitz, J. W.: Int. J. Quant. Chem. Symp. *11*, 501 (1977)
155. Ortega-Blake, I., Leś, A.: Int. J. Quant. Chem. *19*, 463 (1981)
156. Ortega-Blake, I., Novaro, O., Leś, A., Rybak, S.: J. Chem. Phys. *76*, 5405 (1982)
157. Ortega-Blake, I., Leś, A., Rybak, S.: J. Theor. Biol. *104*, 571 (1983)
158. Umeyama, H., Nakagawa, S., Nomoto, T., Moriguchi, I.: Chem. Pharm. Bull. *28*, 745 (1980)
159. Umeyama, H., Nomoyo, T.: ibid *27*, 2504 (1979)
160. Bartsch, R. A., Carsky, P.: J. Org. Chem. *45*, 4782 (1980)
161. Clementi, E.: Lecture Notes in Chemistry, Vol. 2, Springer-Verlag, Berlin 1976
162. Clementi, E.: ibid, Vol. 19, 1980
163. Clementi, E., Corongiu, G.: J. Biol. Phys. *11*, 33 (1983)

164. Clementi, E., in: Structure and Dynamics: Nucleic Acids and Proteins (eds. Clementi, E., Sarma, R. H.), Adenine Press, 1983, p. 321
165. Kim, K. S., Vercauteren, D. P., Welti, M., Chin, S., Clementi, E.: Biophys. J. *47*, 327 (1985)
166. Kim, K. S., Vercauteren, D. P., Welti, M., Fornili, S. L., Clementi, E.: IBM Technical Report Pok.-42, April 20, 1984
167. Corongiu, G., Clementi, E., Pretsch, E., Simon, W.: J. Chem. Phys. *70*, 1266 (1979)
168. Portmann, P., Pretsch, E., Simon, W.: in preparation
169. Pretsch, E., Bendl, J., Portmann, P., Welti, M., in: Proceedings of the Symposium on Steric Effects in Biomolecules (ed. Naray-Szabo, G.), Elsevier-Akademiai Kiado, 1982, p. 85
170. Groth, P., Acta Chem. Scand. *A 36*, 109 (1982)
171. Dobler, M., Dunitz, J. D., Seiler, P.: Acta Crystallogr. *B 30*, 2741 (1974)
172. Seiler, P., Dobler, M., Dunitz, J. D.: ibid *B 30*, 2744 (1974)
173. Nagano, O., Kobayashi, A., Sasaki, Y.: Bull. Chem. Soc. Japan *51*, 790 (1978)
174. a) Czugler, M., Weber, E.: J. Chem. Soc., Chem. Commun. 472 (1981)
b) Weber, E.: J. Org. Chem. *47*, 3478 (1982)
175. Kim, K. S., Clementi, E.: J. Am. Chem. Soc. *107*, 227 (1985)
176. Corongiu, G., Fornili, S. L., Clementi, E.: Int. J. Quant. Chem. Quant. Biol. Symp. *10*, 277 (1983)
177. Schuster, P.: Angew. Chem. *93*, 532 (1981); Angew. Chem., Int. Ed. Engl.
178. Jeziorski, B., Kołos, W.: Int. J. Quant. Chem. *91*, 12 (1977)
179. Dreyfus, M., Pullman, A.: Theor. Chim. Acta *19*, 20 (1970)
180. Kitaura, K., Marokuma, K.: Int. J. Quant. Chem. *10*, 325 (1976)
181. Gresh, N., Claverie, P., Pullman, A.: Int. J. Quant. Chem. Symp. *13*, 243 (1979)
182. Dreyfus, M., Pullman, A.: C. R. Acad. Sci. Ser. C *271*, 457 (1970)
183. Pullman, A., Zakrzewska, K., Perahia, D.: Int. J. Quant. Chem. *16*, 395 (1979)
184. Gresh, N., Etchebest, C., de la Luz Rojas, O., Pullman. A.: Int. J. Quant. Chem. Quant. Biol. Symp. *8*, 109 (1981)
185. Gresh, N., Pullman, A.: Int. J. Quant. Chem. *22*, 709 (1982)
186. Gresh, N., Pullman, B.: Biochim. Biophys. Acta *625*, 356 (1980)
187. Gresh, N., Pullman, B.: ibid *608*, 47 (1980)
188. Gresh, N.: ibid *597*, 345 (1980)
189. Pullman, A., Etchebest, C.: FEBS Letters *163*, 199 (1983)
190. Etchebest, C., Pullman, A.: ibid *170*, 191 (1984)
191. Etchebest, C., Ranganathan, S., Pullman, A.: ibid *173*, 301 (1984)
192. Urry, D. W., Venkatachalam, C. M., Prasad, K. V., Bradley, R. J., Parenti-Castelli, G., Lenaz, G.: Int. J. Quant. Chem. Quant. Biol. Symp. *8*, 385 (1981)
193. Venkatachalam, C. M., Urry, D. W.: J. Comp. Chem. *4*, 461 (1983)
194. Gresh, N., Claverie, P., Pullman, A.: Theor. Chim. Acta *66*, 1 (1984)
195. Diner, S., Malrien, J. P., Claverie, P.: ibid *13*, 1 (1969)
196. Pullman, B., Courriere, P., in: Proc. Jerusalem Symposium Quant. Chem. Biochem. (eds. Bergmann, E. D., Pullman, B.) Vol. 5, p. 547, 1972
197. Bendl, J., Pretsch, E.: J. Comp. Chem. *3*, 580 (1982)
198. Melberg, S., Rasmussen, K.: Carbohydr. Res. *69*, 27 (1979)
199. Melberg, S., Rasmussen, K.: ibid *71*, 25 (1979)
200. Palla, P., Petrongolo, C., Tomasi, J.: J. Phys. Chem. *84*, 435 (1980)
201. Laurence, P. R., Thomson, C.: Theor. Chim. Acta *58*, 121 (1981)
202. Rasmussen, K., Tosi, C.: Acta Chem. Scand. *A 37*, 79 (1983)
203. Burkert, U., Allinger, N. L.: Molecular Mechanics, ACS Monograph 177, American Chemical Society, Washington 1982
204. Ermer, O.: Aspekte von Kraftfeldrechnungen, Wolfgang Bauer Verlag, München 1981
205. Rasmussen, K., Lecture Notes in Chemistry, Vol. 37, Springer-Verlag, Berlin 1985
206. Bovill, M. J., Chadwick, D. J., Sutherland, I. O.: J. Chem. Soc., Perkin II, 1529 (1980)
207. Wipff, G., Weiner, P., Kollman, P.: J. Am. Chem. Soc. *104*, 3249 (1982)
208. Dunitz, J. D., Seiler, P.: Acta Crystallogr. *B 30*, 2739 (1974)

Design of Biospecific Compounds which Simulate Enzyme-Substrate Interaction

Kazutaka Tanizawa and Yuichi Kanaoka

Department of Synthetic Chemistry, Faculty of Pharmaceutical Sciences, Hokkaido University, Sapporo 060, Japan

Table of Contents

This present article surveys the recent development of biospecific compounds which interact with active sites of enzymes. These compounds are classified according to their mode of interaction. The characteristic features of interaction are discussed and the molecular basis for the design of the specific compounds of each type is considered. Significance of the enzyme-specific compounds in basic research and in the application of chemotherapeutics is exemplified. The development of "inverse substrates", specific compounds for trypsin and trypsin-like enzymes of a new type, is also described. The basic idea for the design of inverse substrates and their applicabilities are discussed.

1 Introduction

The most characteristic properties of enzymes which distinguish them from other chemical catalysts are those associated with their specificity. It is well recognized that binding of a substrate to an enzyme takes place at an active site containing the catalytic function and that formation of an enzyme-substrate complex always precedes the catalytic process. Therefore, enzyme-substrate interaction is generally realized to be one of the most crucial processes in the sense that accurate molecular recognition is involved.

There are, however, many substances other than physiological substrates which exhibit specific interactions with the active site of enzymes. They include small, synthetic competitive inhibitor molecules, synthetic quasi-substrates, affinity labeling reagents, mechanism-based inhibitors, and so on. This may suggest that these substances can act, because enzymes exhibit some structural allowance in substrate recognition. These specific compounds are expected to be of great value for application in basic research and also in the medicinal field. Some of these specific compounds have reached clinical uses.

This review will deal with the design of specific compounds and their applications, mainly concerning hydrolytic enzymes, with which a variety of studies have been carried out. The development of a new type of specific substrates for trypsin and trypsin-like enzymes is also described.

2 Classification of Enzyme-Specific Compounds

Compounds which exhibit specific interactions with a particular site of an enzyme other than an active site are called cofactors and allosteric effectors. These compounds are not considered in this review. Specific compounds which interact with the active site itself will be classified into two types. One of them include simple competitive inhibitors and photoaffinity labeling reagents. Compounds of this type exhibit a specific interaction only with the binding site (specificity site) of the enzyme.

Table 1. Classification of enzyme-specific compounds

enzyme interacting site		specific synthetic compound
active site	binding	competitive inhibitor, photo-affinity labeling reagent
	binding and catalytic site	transition state analog, quasi-substrate, mechanism-based inhibitor, affinity labeling reagent
other than active site	cofactor binding site	chemical modification reagent
	allosteric site	chemical modification reagent, synthetic allosteric effector

Compounds of the other type are those which interact with both the binding and the catalytic site. Substrate analogs of various types and specific irreversible inhibitors show this behaviour. Compounds of the latter category are the main subject of this review, while compounds of the former type are discussed only briefly.

3 Design of Binding Site-Interacting Substances

Large numbers of competitive inhibitors for a variety of enzymes have been reported. The design of inhibitors does not pose difficulties, as the site-specific group for the enzyme binding site is the only parameter for the molecular design. This group will determine the intermolecular forces and the spatial adaptation exhibited between enzyme active site and inhibitor. The forces involved in the binding are noncovalent: they may be relatively strong forces such as electrostatic interactions and hydrogen bonds, and also weaker contributions from hydrophobic bonding and van der Waals or London dispersion forces. For a good inhibitor, the cumulative effects of such forces produce tight binding at the active site. The practical method of design, however, has been largely empirical. Screening of a large number of analogs derived from the lead compound of known activity is one of the effective ways to develop potent compounds. These data also serve to predict new highly effective compounds in a statistical methodology — Quantitative Structure Activity Relationship (QSAR). Recently, computer-assisted drug design has been paid much attention [1]. This method will become increasingly important in future.

3.1 Applications in Research

Affinity chromatography, based on biological recognition, has become a major means for the purification of biologically active molecules [2]. The technique provides a simple and effective way of purification. Specific adsorption of the enzyme to its competitive inhibitor attached to a polymer matrix is the basis for an efficient enzyme purification. Affinity electrophoresis is also based on biological recognition[3].

Photoaffinity labeling reagents can be regarded as substances involved only in binding site interactions. The reagents include both site-specific groups and potentially reactive groups (photo-reactive groups), and the reagents themselves are simple competitive inhibitors. The photo-reactive group is not necessarily designed to aim at the catalytic functional group of the enzyme molecule. Rather, the reagents have a unique significance in mapping active site structure, because, e.g., carbens and nitrens once generated by photolysis are highly reactive and indiscriminately so towards a variety of amino acid residues near the binding site [4].

3.2 Applications in Medicinal Fields

Competitive inhibition involves simple processes as shown in Eq. (1). Many competitive inhibitors are practically used as therapeutic agents, though they have to be qualified in many aspects. It is an undeniable fact that even a potent inhibitor *in vitro* does not always prove satisfactory *in vivo*. There are many factors affecting the efficiency of enzyme inhibitors in the living system. The existence of the natural substrate is one of the most important points *in vivo*. The degree of inhibition is directly related to the ratio of the inhibitor concentration (I) divided by its inhibition

$$\begin{array}{l} E + I \overset{K_i}{\rightleftharpoons} EI \\ + \\ S \\ \Big\downarrow\!\!\uparrow K_m \\ ES \\ \downarrow \\ P \end{array} \tag{1}$$

constant (K_i) to the substrate concentration (S) divided by the Michaelis-Menten constant (K_m). Therefore, for a very potent inhibitor the molecular design is important. A strong binding affinity of an inhibitor also serves to decrease the dose amount, and this simultaneously prevents undesired non-specific effects on untargeted enzymes.

Captopril is one of the well-known examples of a competitive inhibitor used as a drug. It has been expected that an inhibitor of the angiotensin-converting enzyme is effective to reduce blood pressure [5]. The design of specific inhibitors of the enzyme followed the structure of its substrate, angiotensin I, and its inhibitor, snake venom. Thus, captopril (D-3-mercapto-2-methylpropanoyl-L-proline) (*1*) is now clinically used as an orally active antihypertensive drug [6]. The estimated interaction between the inhibitor and the active site is shown in Fig. 1. The binding affinity could be produced by electrostatic interaction, hydrogen bonding, and hydrophobic subsite interaction, and no catalytic residues participate in the binding process. The K_i and IC_{50} values are reported to be as small as 1.7×10^{-9} M and 2.2×10^{-8} M, respectively and did not inhibit most other peptidases until added at a concentration of 10^{-3} M [7]. Several new inhibitors modeled after captopril have been reported. Zofenopril (*2*) [8], SA-446 (*3*) and its benzoyl analog (*4*) [10] are recognized to be more potent and exhibiting longer activity than captopril. The lactam (*5*) (half as active as captopril) was developed using a computer-assisted molecular modeling approach [11].

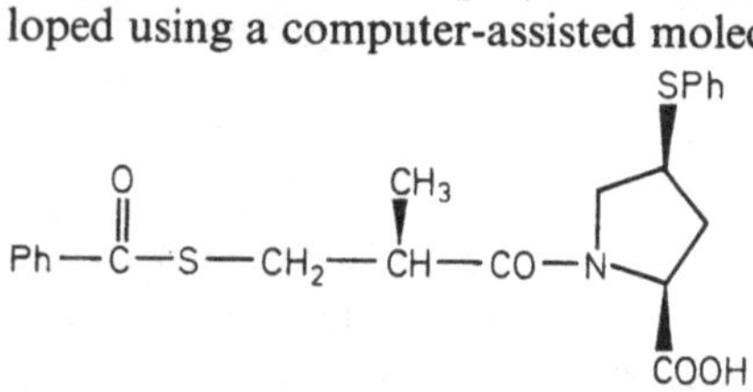

2

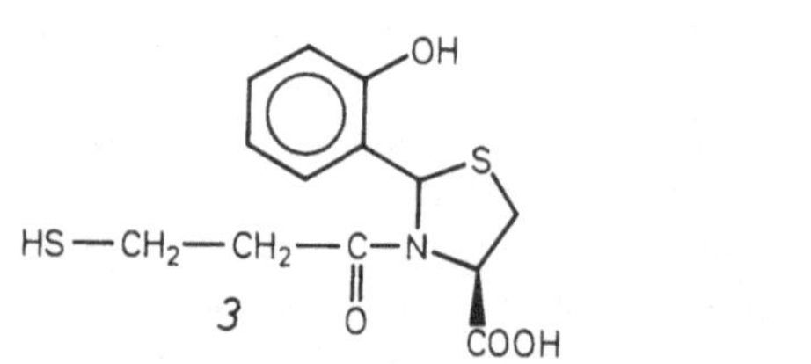

3

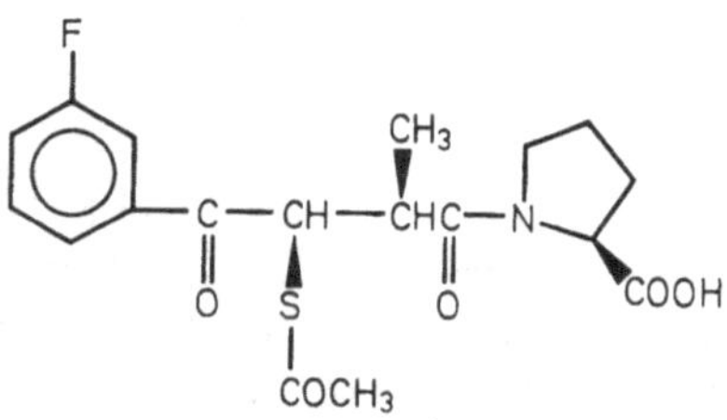

4

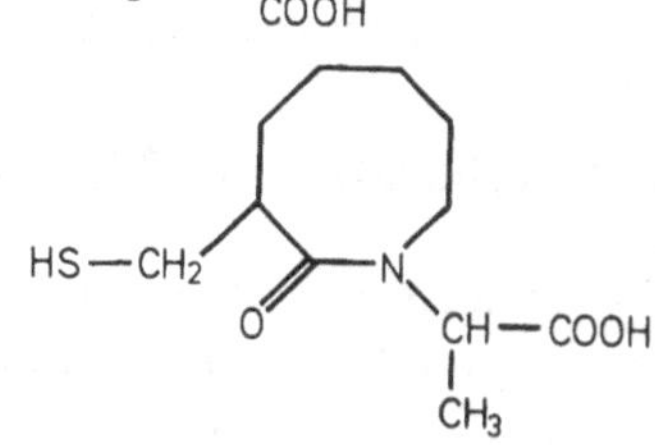

5

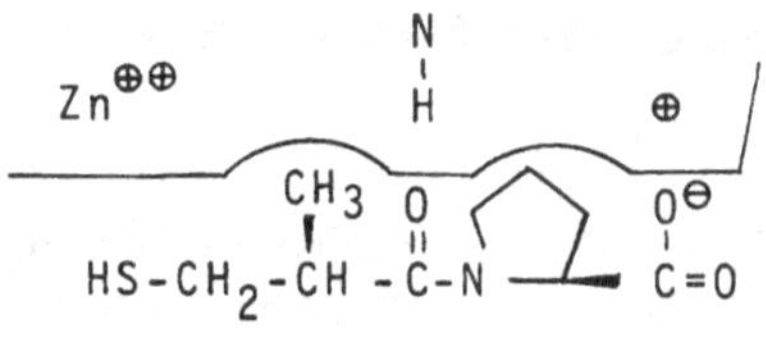

1

Fig. 1. A hypothetical model for the binding of captopril to the active site of an angiotensin-converting enzyme

trans-4-(Aminomethyl)cyclohexanecarboxylic acid is another example of a competitive inhibitor used as a drug. It was developed on the basis of the structure of plasmin inhibitors, ε-aminocaproic acid and lysine. It is clinically applied as an antihemorrhagic agent [12].

The most important requirement for the chemotherapeutic agents is their selective toxicity. The most ideal agents are those which are directed toward enzymes of foreign pathogens or aberrant cells (cancer) without affecting host enzymes. The selectivity to be exhibited *in vivo* would be difficult to predict solely from *in vitro* data. In this case, however, chemotherapeutic agents are ideal since the target enzyme of the pathogen has no counterpart in the host and the inhibitor is target-specific. This situation is approximated with β-lactam antibiotics (cf. Sect. 4.5). On the other hand, the presence of a homologous enzyme in the host does by no means preclude selectivity, as demonstrated by the very useful antibacterial agent, trimethoprim. Trimethoprim (*6a*), an analog of dihydrofolic acid (*7*), acts as a dihydrofolate reductase inhibitor and exerts its effect simply by binding. Selective toxicity in this case is fortunately exhibited by a large difference in inhibitor specificity, i.e., the K_i value for the bacterial enzyme is several thousand times lower than that of the host enzyme [13]. The exemplified case of trimethoprim suggests that the development of a drug is often attained empirically. There are a variety of *in vivo* factors such as pathogen-host relationships which determine whether the enzyme inhibitors are practically useful for clinical purposes. Our knowledge to predict the effects of *in vivo* factors on the performance of enzyme inhibitors is still limited. Rational approaches in the chemotherapeutic field should evolve parallel with our knowledge of comparative biochemistry and metabolism. Recently, the design of drugs to fit macromolecular receptors, including enzymes, has attracted much attention [14]. Molecular modeling considerations of dihydrofolate reductase and trimethoprim derivatives led to the replacement of one *meta*-methoxy group of *6a* by a carboxyalkyloxy group, and, furthermore, a chain length was selected to optimize the interaction between the carboxylate and the guanidinium group of Arg-57 of the enzyme. Compound *6b* was found a much more potent inhibitor than the original trimethoprim [15,16].

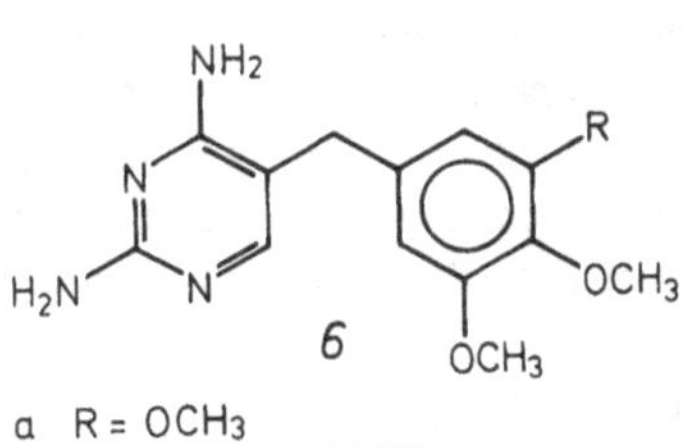

6

a R = OCH_3
b R = $O(CH_2)_5COOH$

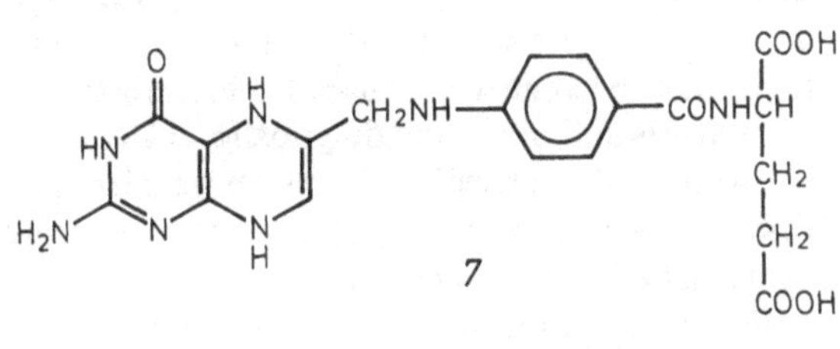

7

4 Design of Specific Substances Involving both Binding Site and Catalytic Site Interactions

Compounds of this type must have a structure with two separate moieties, the binding site partner and the catalytic site partner, being spatially arranged according to the active site structure of the enzyme. Enzymatic processes involved in the interaction with such compounds are shown in the following equations:

synthetic and quasi-substrates:

$$E + S \overset{K_i}{\rightleftharpoons} ES \xrightarrow{k_1} E + P \tag{2}$$

transition state analogy:

$$E + I \overset{K_i}{\rightleftharpoons} EI \underset{k_{-1}}{\overset{k_1}{\rightleftharpoons}} [E \ldots I'] \tag{3}$$

irreversible affinity labeling inhibitor:

$$E + I \overset{K_i}{\rightleftharpoons} EI \xrightarrow{k_1} E - I' \tag{4}$$

mechanism-based inhibitor:

$$E + I \overset{K_i}{\rightleftharpoons} EI \rightleftharpoons [E \ldots I'] \begin{cases} \xrightarrow{k_2} E - I'' \\ \xrightarrow{k_3} E + P \end{cases} \tag{5}$$

inhibition by a stable intermediate:

$$E + S \overset{K_s}{\rightleftharpoons} ES \xrightarrow{k_1} E - A \xrightarrow{k_2} E + P \tag{6}$$

4.1 Synthetics and "Quasi"-Substrates

A variety of synthetic substrates for a variety of enzymes have been reported. Especially for hydrolytic enzymes, many substrates have been prepared owing to the ease of their design. Chromogenic and fluorogenic substrates are of special value for simple and sensitive spectrometric determinations of enzyme activities. Thus these compounds lately have become widely used for investigations of various proteases both in research laboratoris and in clinical diagnostics. Chromogenic [17] and fluorogenic [18,19] peptidyl substrates interact with subsites and exhibit pronounced specificity. They are useful for the specific detection of a certain protease from a sample containing several proteases of similar specificity.

For determining the absolute concentration of active hydrolytic enzymes, active site titrants, a sort of quasi-substrates, have been developed. The catalytic pathway of hydrolytic enzymes involves an acyl enzyme intermediate as shown in Eq. (7), which is equivalent to Eq. (6).

$$E + S \underset{}{\overset{K_s}{\rightleftharpoons}} ES \xrightarrow{k_1} E - A \xrightarrow{k_2} P_2 \quad (\searrow P_1) \tag{7}$$

Therefore the compounds designed to give a stable acyl enzyme intermediate ($E - A$) in a specific manner with a concomitant release of spectrometrically detectable product (P_1) are useful as active site titrants of an enzyme[20-22]. Compounds *8* [23] and *9*[24] are proposed as titrants for trypsin. Both have a site-specific residue for trypsin, the guanidinophenyl or amidinophenyl moiety, as well as a chromogenic leaving portion, though they lack an α-acylamido group and an asymmetric carbon atom which are basic constituents of natural and synthetic substrates. These structural characteristics realize a favorable acylation subsequently to the specific binding. The deacylation, however, was shown to be much slower than the acylation step. The kinetic properties are advantageous for the titrant. The release of a stoichiometric amount of *p*-nitrophenolate (P_1) is monitored by optical density at 405 nm. Kinetic parameters for *8* and *9* are listed in Table 2. The use of a fluorophore instead of a

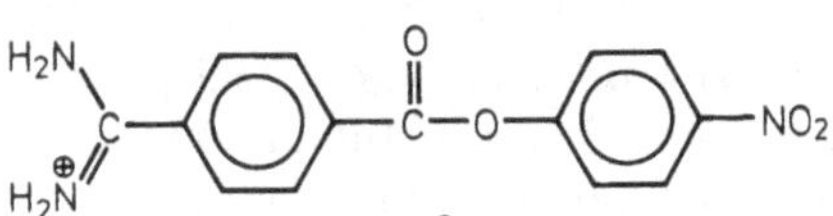

8

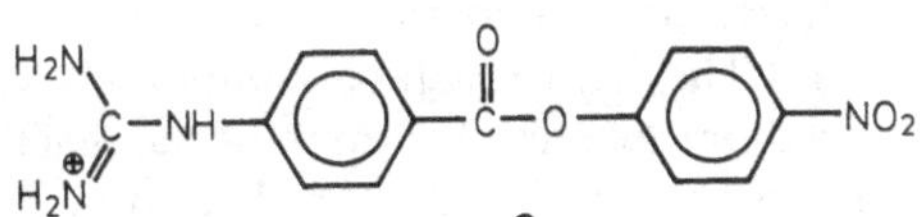

9

chromophore in the design of substrates and titrants provides an increase in sensitivity. Fluorogenic active site titrants for trypsin and trypsin-like enzymes *10* [20] and *11* [25] were designed following compound *9*. It was reported that the detectability of the enzyme concentration was increased 5—6 orders of magnitude, from 10^{-6} M for *9* to $10^{-11} - 10^{-12}$ M for *11*.

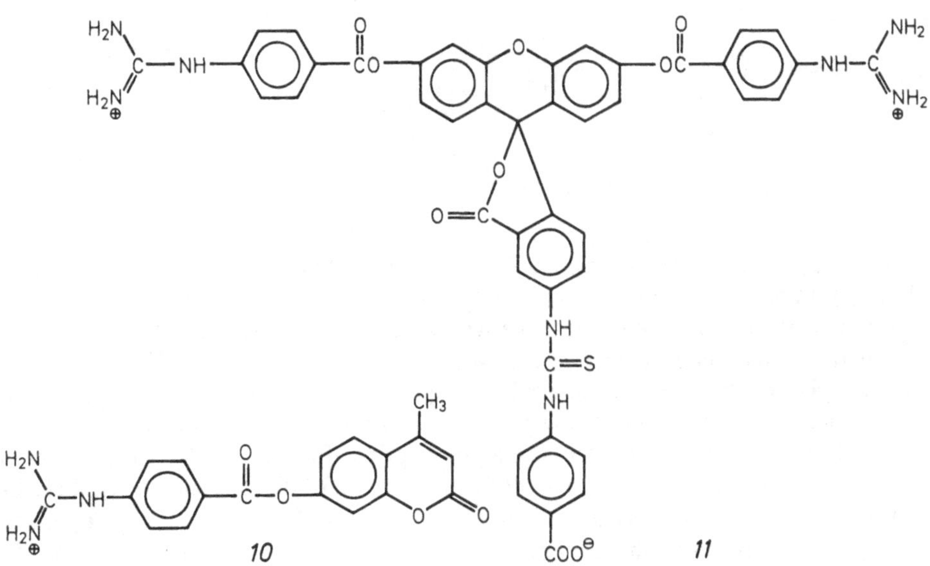

10 *11*

Table 2. Kinetic parameters for trypsin-catalyzed hydrolysis of *8* and *9*

Compound	K_s, M ($\times 10^5$)	k_2, s^{-1}	k_3, s^{-1} ($\times 10^4$)	pH
8	0.503	30.4	653	8.2
9	0.061	1.95	0.34	8.3

4.2 Transition State Analogs

It has been speculated that the catalytic specificity of an enzyme requires the active site of the enzyme and the transition state of the reaction at the substrate molecule to be structurally complementary [26]. Molecules which resembled the transition state structure could thus be expected to bind the active site tightly. This concept was taken up and developed by Lienhard [27] and Wolfenden [28] as transition state analogs.

The design of a transition state analog is based on knowledge of the mechanism of the target enzyme. Enzyme mechanisms which involve a change in bond order, i.e., trigonal-tetrahedral or tetrahedral-trigonal transformations, are most suited. These mechanisms are found in a number of reactions such as hydrolysis and transfer reactions. The design of stable compounds which mimic the transition state is generally carried out by a modification of the reacting functional group of its common substrate. The proper choice of an alternative for the reacting functional group is important to produce the resembled transition state. It should be noted that the inhibition mode of the transition state analog is competitive, though its K_i value is generally much smaller than that of a simple competitive inhibitor. Typical examples of this approach are boronic acid inhibitors for hydrolases. Some of the hydrolases catalyze acyl transfer reactions via the intermediacy of an ester with a seryl residue within the active site. The transition states for the acylation and deacylation steps of these enzymes are thought to involve a metastable tetrahedral intermediate (*12*).

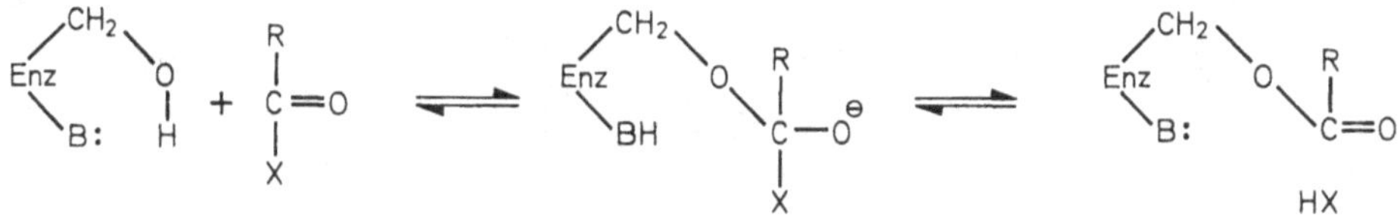

12

Boronic acid derivatives form stable tetrahedral adducts with hydroxide ion and they behave as strong inhibitors of hydrolases. This leads to the assumption that the boronic acid derivatives bind to the serine residue at the active site of the enzymes in a structure resembling the tetrahedral intermediate (*13*) [29]. The binding affinity of N-benzoylaminomethaneboronic acid for chymotrypsin, for example, is reported to be two orders of magnitude stronger than that of a hippuric acid derivative [30].

Another example are naturally occurring peptide aldehyde inhibitors, discovered in microorganisms, such as antipain (*14*), chymostatin (*15*), leupeptin (*16*) and elastinal (*17*) [31–33]. The discovery of the inhibitors stimulated the synthetic work of peptide aldehyde analogs, and a large number of peptide aldehydes have been prepared [34].

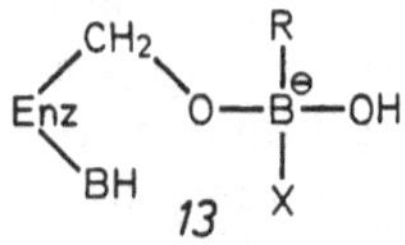

13

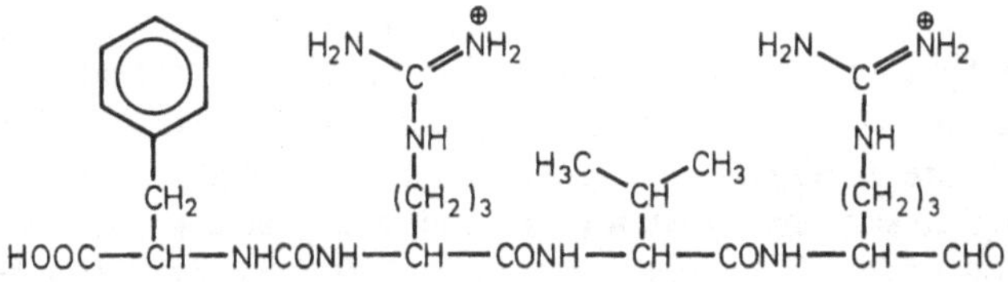

14

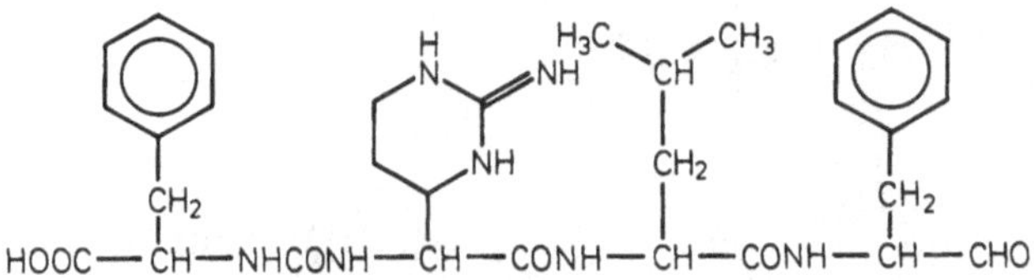

15

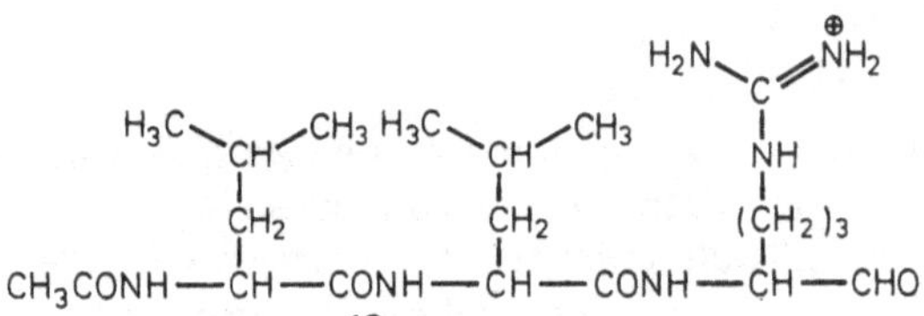

16

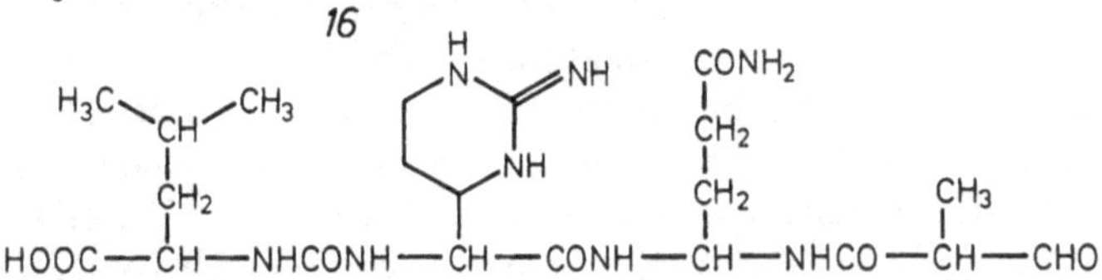

17

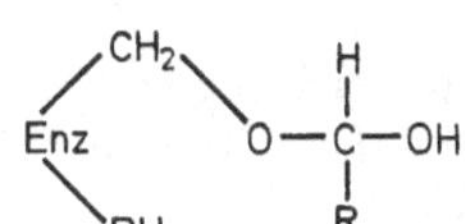

18

The formation of a tetrahedral hemiacetal adduct was analyzed for the interaction between the inhibitor aldehyde and the catalytic serine residue (*18*) [35]. The overall dissociation constants for an enzyme and an interacting transition state analog may be given by:

$$K_{i(overall)} = K_i \times k_{-1}/(k_1 + k_{-1}) \tag{8}$$

The apparent strong affinity is reasonably assumed to arise from a second equilibrium step where k_{-1} is much smaller than k_1. A kinetic analysis of the leupeptin-trypsin interaction revealed that the dissociation constant for the entire process ($K_{i\,\text{overall}}$) is 1.34×10^{-8} M, though that of the first step (K_i) is only 1.24×10^{-3} M. The contribution of the second equilibrium to the entire process was determined to be a magnitude of 10^5 [36].

Trimethylammonium trifluoroacetophenone (*19*) was found to be a highly effective inhibitor of acetylcholinesterase [37]. The ketone activated by an electron-withdrawing trifluoroacetyl group will enhance the tendency to add a nucleophile (the hydroxyl group of the catalytic serine residue of acetylcholinesterase) to form a tetrahedral adduct as an aldehyde inhibitor.

p-Amidinophenylpyruvic acid (*p*-APPA, *20*) was first discovered by Geratz to be a good trypsin inhibitor [38]. Spectrometric analysis of the interaction of thrombin and trypsin with *p*-APPA led to the conclusion that the excellent inhibitory properties of *p*-APPA are explained by a transition state mechanism: formation of a hemiketal complex. In contrast, *m*-amidinophenylpyruvic acid (*m*-APPA) which is apparently incapable of forming a hemiketal, did not afford any evidence for the formation of a hemiketal complex and displayed a K_i in the range expected for simple benzamidine [39].

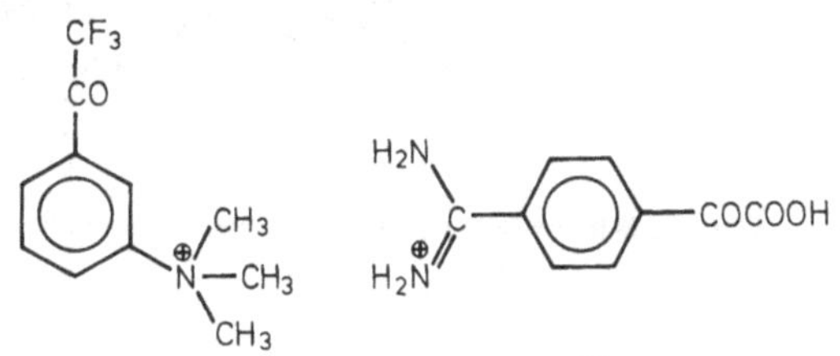

19 20

The X-ray diffraction experiment on the lysozyme-inhibitor complex[40] is a well-known example which gave evidence for the transition state complementarity.

Lysozyme catalyzes the hydrolysis of cell wall and synthetic polymers of β(1–4)-linked units of N-acetylglucosamine (NAG). During the catalysis, it is expected that a carbonium ion is formed in which the conformation of the glucopyranose ring changes from full-chair to a half-chair conformation. This speculated conformation is in accordance with the X-ray data. Thus, the designed transition state analog (*21*), in which the lactone ring mimics the carbonium ion-like transition state (*22*), binds tightly to lysozyme. The K_i value, 8.3×10^{-8} M, is compared with that for $(NAG)_4$, 10^{-5} M [41].

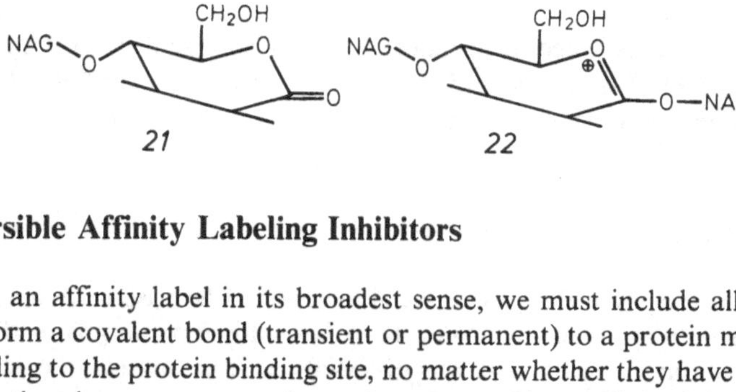

21 22

4.3 Irreversible Affinity Labeling Inhibitors

Considering an affinity label in its broadest sense, we must include all compounds which can form a covalent bond (transient or permanent) to a protein molecule after specific binding to the protein binding site, no matter whether they have a potentially reactive functional group or actual reactive group. Photoaffinity labeling reagents (see Sect. 3.1) and transition state analogs (Sect. 4.2) as well as further compounds to be mentioned in the following sections (4.4–4.5) should be included here. At this point, however, we will deal only with the affinity labeling inhibitors in the narrowest sense which lead to "permanent" inactivation associated with a chemical modification by the "actually" reactive functional group at the active site.

Covalent bond formation is principally considered the most effective way to inactivate an enzyme. The inactivation process is time-dependent and the rate depends on the k_1 value. The specificity of the inactivation reaction is dependent on the K_i value; a large K_i value significantly retards the inactivation rate especially for the reaction *in vivo* where the specific compounds, such as the physiological substrate, are present. A variety of classes of chemical reactions can be used to modify the enzyme. These are: nucleophilic substitution, nucleophilic addition, electrophilic substitution, etc. The selection of the reactive group depends on the target functional group of the active site to be modified. The design of a reactive group with a very enhanced reactivity is considered to be unsuitable, because random reactions with the enzyme surface as well as reactions with solvent water will take place. It may be noted that affinity labeling itself exhibits enhanced reactivity through the proper orientation of the reactive group of the inhibitor to the functional group of the enzyme. Therefore the selection of a group with a rather diminished reactivity may be preferable. In addition, a reactive group with a wide reactivity-spectrum is advantageous for the purpose of topographical mapping of the enzyme active site in which only amino acid residues close to the inhibitor are modified. A nucleophilic substitution reaction is best suited for this purpose, because many amino acids have a nucleophilic group in their side chain.

α-Haloketones are one of the most popular chemical classes of affinity labels. Since haloketones are reactive with most nucleophiles, they have a good chance to modify the closest located residue. Typical examples of this class are chloromethylketones derived from N-tosyl-L-phenylalanine (TPCK) (*23*) and N-tosyl-L-lysine (TLCK) (*24*)[42]. They react with the histidine side chain in the catalytic site of chymotrypsin and trypsin, respectively. α-Haloketone (*25*)[42] in combination with a guanidinophenyl moiety, an efficient ligand for trypsin, results in alkylation of a serine residue at the catalytic site. The different behavior of *24* and *25* on the reacting residue of the enzyme active site reflects the geometry of the active site complementary to the reagents; only the nucleophile which comes close to the reactive group of the inhibitor during EI complex formation is involved in the modification. The applicability of affinity labels with chloromethylketone was further extended by the development of peptide chloromethylketones which incorporate a part of the sequence of the physiolo-

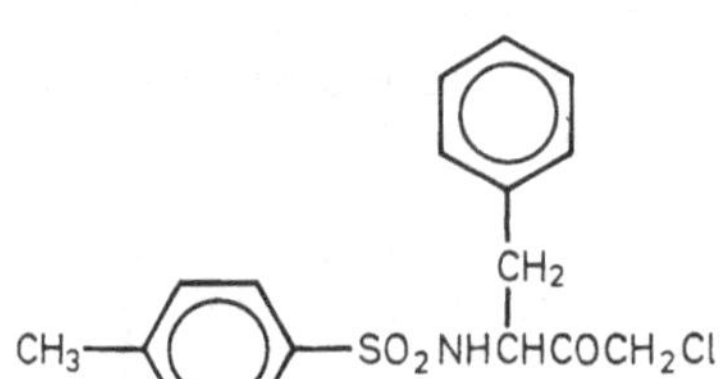

23

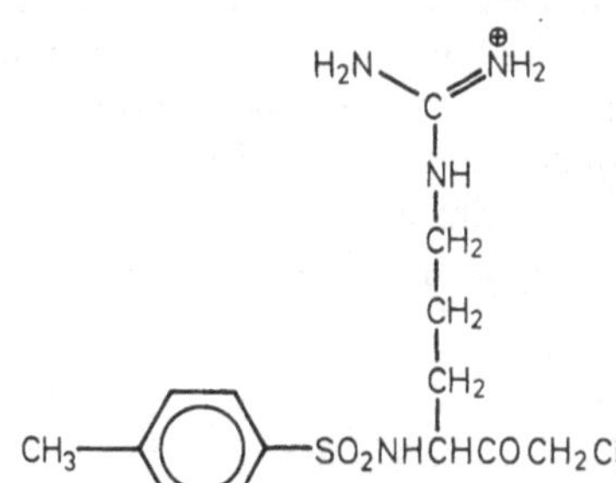

24

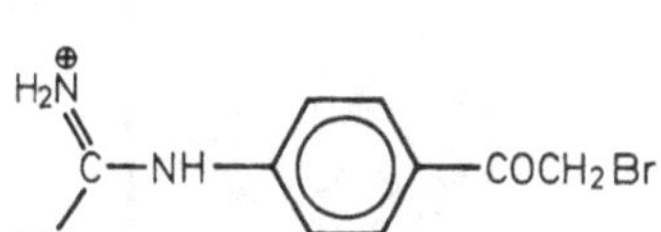

25

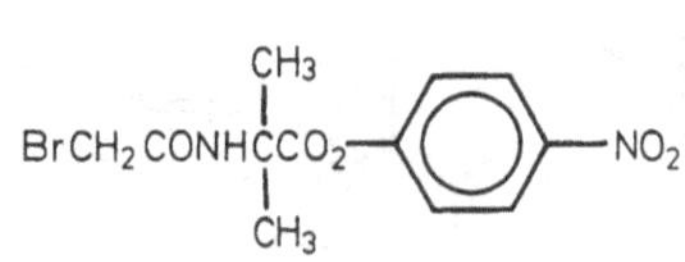

26

gical substrates of enzymes. These reagents distinguish among proteases of similar specificity by taking advantage of binding selectivity in both primary and secondary sites [43]. α-Haloacetamide is another typical example with a wide range of reactivity. Modifications of carboxypeptidase B by bromoacetyl reagents with different structure are highly diverse. Alkylated residues of, e.g., tyrosine, glutamic acid, or methionine were changed by such reagents residues [44-46]. Sulfonyl halide and diazonium are also useful for the reactive group of affinity labels.

Efficient modification steps through the proper orientation of the inhibitor reactive group to the enzyme nucleophile is realized by covalent bond formation. A classic example of this type is the modification of a methionine residue of chymotrypsin by *p*-nitrophenyl bromoacetyl α-aminoisobutyrate (*26*) [47]. In this instance, the reactive group (bromoacetyl) is fixed at the locus near the active site through a covalent bond by means of acyl enzyme intermediates.

4.4 Mechanism-Based Irreversible Inhibitors

Different names have been used by several reviewers for this type of compounds: suicide enzyme inhibitors, suicide substrates, k_{cat} inhibitors, Trojan horse inhibitors, etc. Compounds of this type are chemically unreactive, but their products from enzymatic conversion are highly reactive molecules. These products, formed within the active site, may immediately attack an essential protein residue or prosthetic group, resulting in the irreversible inhibition of the enzyme. Thus, the specificity of the inhibitor is determined not only by the binding affinity but also by its effectiveness to serve as a substrate for the target enzyme. Inhibitiors of this type are more specific than simple affinity labeling reagents, because they are chemically unreactive to foreign biomolecules. Requirements for the design of mechanism-based irreversible inhibitors are very stringent; .

1) the molecule must be chemically unreactive;
2) it must behave as a specific substrate for the target enzyme; and
3) the resulting product must be spatially well-arranged and active enough to react with an active site residue of the enzyme without being quenched by the solvent.

The third requirement is not absolute, as a stoichiometric inactivation will occur if the condition $[E] \ll [I]$ applies. Even if modifications occur less-frequently than in quenching, the modification can be ultimately completed after several catalytic turnovers. Since an excellent review has recently appeared on mechanism-based enzyme inactivators [48], we will mention a few selected examples here.

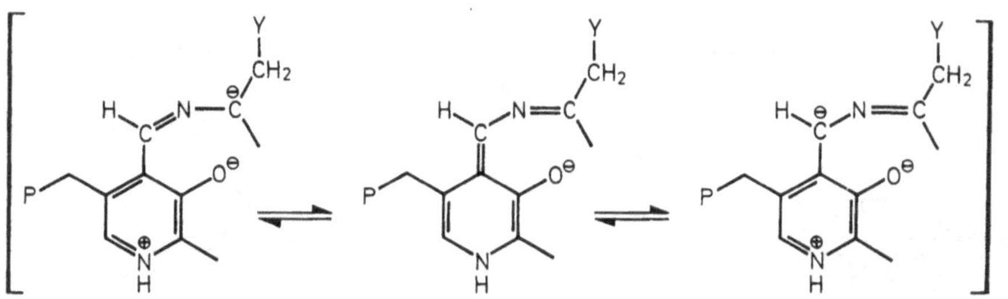

27

Enzymes requiring pyridoxal phosphate (PLP) have been prime targets for the design of mechanism-based inhibitors [49]. The coenzyme pyridoxal phosphate condenses with amino acids to form a Schiff base. The pyridine ring in the Schiff base acts as an electron sink which effectively stabilizes a negative charge. Each one of the groups around the α-carbon may be cleaved, forming an anion which is stabilized by the Schiff base and the pyridine ring. For example, the breaking of the α-hydrogen gives a stabilized α-carbanion (*27*) which may react in several different ways. The many examples of mechanism-based inactivation of pyridoxal phosphate-dependent enzymes can be explained in terms of alterations in the fate of the analogous intermediate *27* brought about by the design of appropriate groups on the substrate. Ethanolamine-O-sulfate, for example, is so designed as to generate highly active Michael acceptors (*28*) [50]. The sulfonate group mimics the carboxyl of γ-aminobutyric acid and serves as a leaving group to generate an unsaturated imine, which can alkylate a basic group of the enzyme. Covalent bond formation can occur between *28* and a nucleophile residue to give *29*.

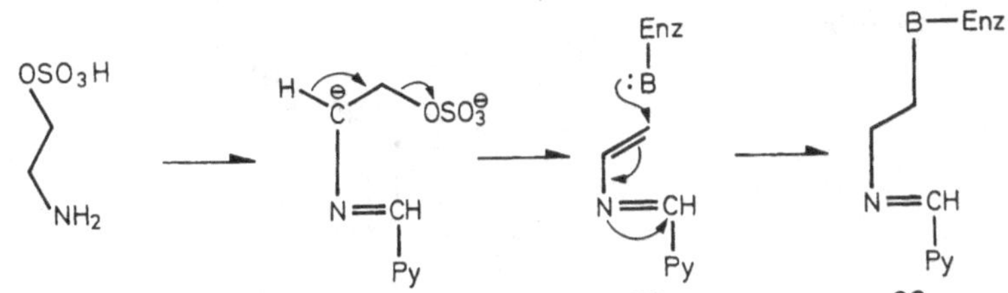

28 *29*

Azaserine, 5-diazo-4-oxo-L-norvaline (DONV) and 6-diazo-5-ketonorleucine (DON) are other examples of mechanism-based irreversible inhibitors [51]. They are stable to nucleophilic attack, but on enzymatic protonation, they are converted to the reactive diazonium ions (*30*). N-Nitroso compounds have been proposed as irreversible inhibitors of proteolytic enzymes. N-Nitrosolactam (*31*) can inhibit chymotrypsin

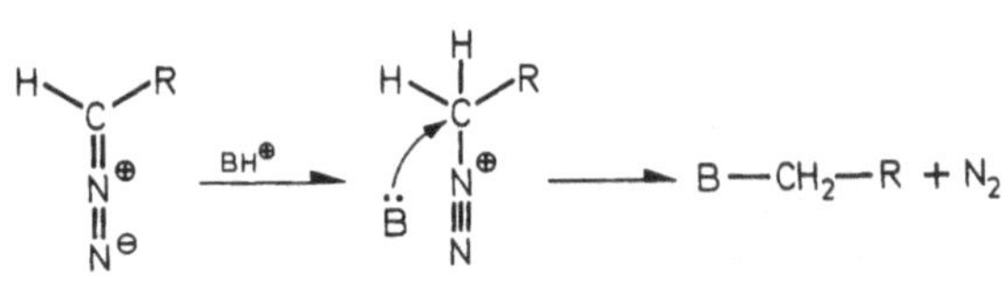

30

irreversibly, possibly by a carbonium ion (*32*) [52]. It is proved from the results that compound (*31*) is eventually well-designed to meet the requirement of the chymotrypsin active center with respect to the specific binding and the proximity to the catalytic residues. Benzylchloropyrone, *33*, inactivates chymotrypsin after 14–40 turnovers. The key step is the enzymatic activation of a latent chloropyrone to an acyl chloride, *34*, during acyl enzyme formation [53]. The reagents enter into a covalent anchoring to the protein particle at the one end of the molecule, just before they are involved in the enzymatic activation, to generate a reactive function at the other end. Another example of an inhibitor for chymotrypsin is 6-iodomethylene-naphthyltetrahydropyrane-2-one (*35*). The compound has a partition ratio of 1.7 turnovers per chymotrypsin inactivation [54]. Labeling reactions by *31*, *33*, and *35* are considered to proceed through an intramolecular process similar to the reaction with *26*.

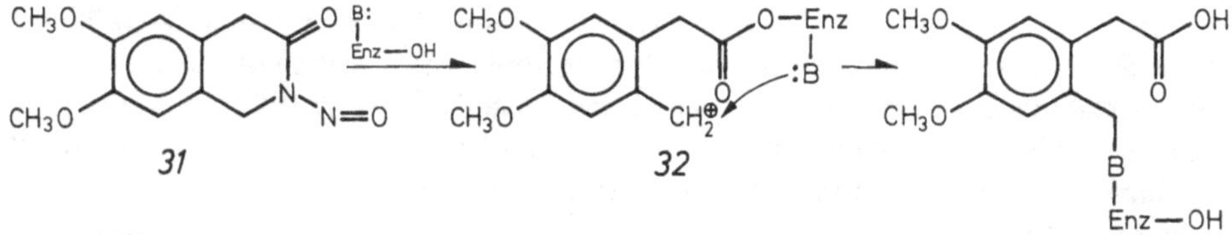

31 *32*

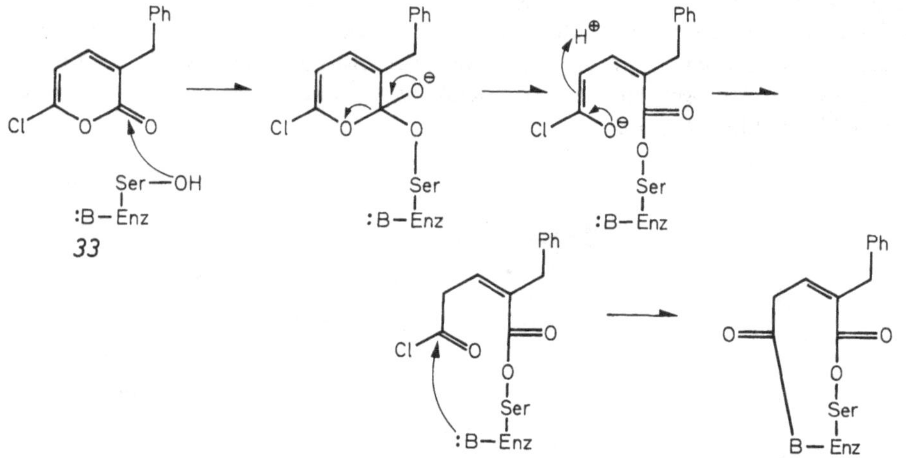

33 *34*

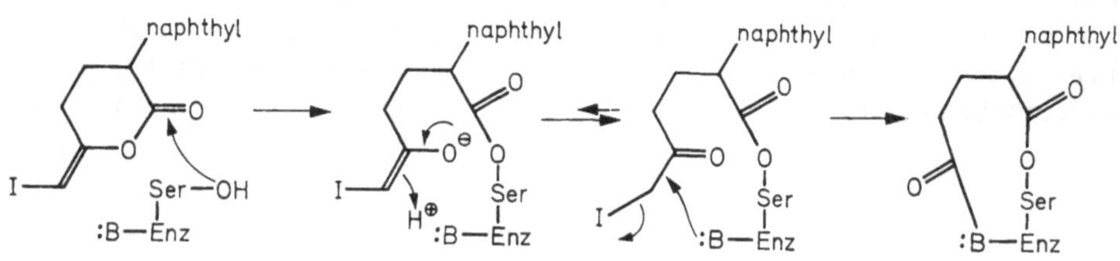

35

4.5 Stable Intermediates

Compounds effecting a stable intermediate in the course of enzymatic catalysis are a sort of mechanism-based inhibitor. However, in this case, the enzymatic activity lost by the formation of the intermediate can regenerate after a certain period. Compounds of this class are often observed for hydrolytic enzymes. The formation of an acyl enzyme intermediate (*EA*) is a characteristic feature of the reaction catalyzed by these enzymes, as shown in Eq. (6). Esters of *p*-guanidinobenzoate (*9*), which were discussed in Sect. 4.1, behave as transient inhibitors of trypsin due to the formation of a relatively stable acyl enzyme. A similar type of inhibition occurs in the temporary inactivation of acetylcholinesterase by the carbamate *36* [55]. An electronically stabilized acyl chymotrypsin was designed by the use of isatoic anhydride (*37*). Isatoic anhydride is sufficiently reactive to give the initial acyl enzyme *38* which transforms into the electron-releasing anthranilyl chymotrypsin *39* upon ready hydrolysis of the carbamate [56]. The electronic nature of the acyl groups govern to a considerable

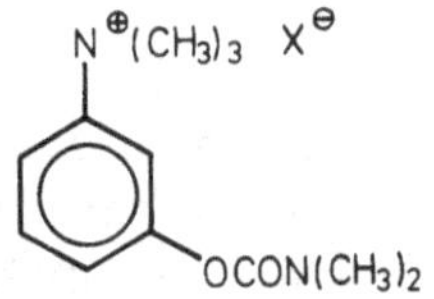

36

extent the stability of the acyl enzyme. An acyl enzymes substituted by an electron-releasing group is generally resistant to hydrolysis though its formation (acylation) step is not favored. Compound *37* is, therefore, designed to satisfy both, the requirement for the efficient production of the acyl enzyme and for the stabilization of the resulting acyl enzyme intermediate. It seems convincing that the electron-releasing character of an acyl group generally enhances the stability of acyl chymotrypsin and acyl acetylcholinesterase toward hydrolysis. In the study of the deacylation rates of *p*-substituted benzoyl enzymes, it was concluded that the rates generally correlate well with the substituent constant (sigma) though there are some exceptional cases. The *p*-guanidino group is one of the exceptional cases to give a very stable acyl trypsin (cf. compound *9*) and an unstable chymotrypsin. The different response of the *p*-guanidinobenzoyl group on the deacylation rates catalyzed by plasmin and thrombin if of interest with regard to their application to the medicinal field [57].

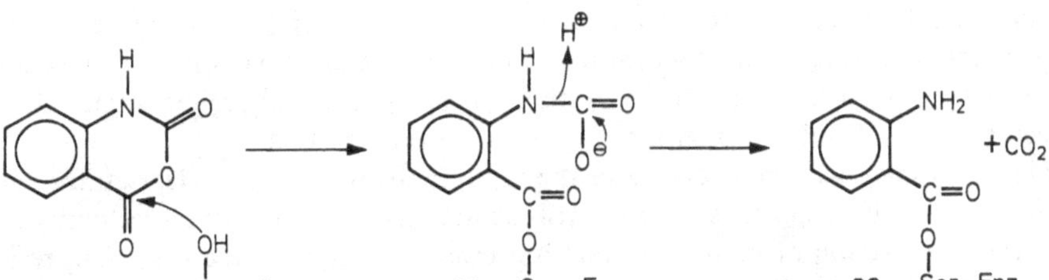

37 *38* *39*

The action of β-lactam antibiotics is considered to be due to the formation of an acyl enzyme with carboxypeptidases and transpeptidases which are involved in the biosynthesis of bacterial cell walls [58]. A three-step mechanism involving a stable acyl-enzyme intermediate (EI*), a participating active site serine residue, and a very slow decay process (k_4) was proposed [Eq. (9)][59].

$$E + I \underset{}{\overset{K}{\rightleftharpoons}} EI \xrightarrow{k_3} EI^* \xrightarrow[\text{nucleophile}]{k_4} E + P \tag{9}$$

Synthetic compounds which afford a stable intermediate must be designed with a structure closely related to the natural substrate of the enzyme. They are obliged to behave as a quasi-substrate in exhibiting specific binding to the binding site and a suitable juxtaposition for the bonding to the catalytic residue. Furthermore, the design must afford an intermediate which is structurally differentiated to remain unchanged for a certain period.

4.6 Applications in Research

Synthetic substrates and titrants have played an important role for understanding the kinetic characteristics and catalytic mechanism of enzymes. For the purpose of enzyme purification, transition state analogs are useful for an affinity ligand in the same manner as a competitive inhibitor, though in some cases difficulties may arise caused by the high affinity to the ligand in the elution process. [60]. The most significant contribution of affinity labels and mechanism-based inhibitors is the elucidation of active site structures and the catalytic mechanism of enzymes. The amino acid analysis of the labeled peptide fragments provides information about the structure of the active site. By use of differently designed affinity labeling reagents, it is possible to determine the spatial outline of the active site as in the case of trypsin-specific chloromethylketone derivatives.

4.7 Applications in Medicinal Fields

In recent years, the importance of enzyme levels in body fluids for clinical diagnosis has been recognized. It has been established that activities of secreted enzymes and cellular enzymes in serum are a sensitive indication of the pathophysiological condition of the body. Specific and sensitive substrates play a prominent role for this purpose. Fluorogenic substrates, e.g., enable sensitive micro-analyses.

In drug design, affinity labeling reagents and transition state analogs are considered potentially promising. Unlike simple competitive inhibitors, transition state analogs and affinity labels appear to offer unique properties by means of additional interaction with the catalytic residues. Mechanism-based inhibitors and stable intermediates are also advantageous because they are essentially inert as chemical reagents until they are specifically activated by the enzyme which is to be modified.

It is known that β-lactamase catalyzes the rapid hydrolysis of the β-lactam ring of penicillins and cepharosporines. The hydrolytic activity of these enzymes eliminates the bacteriocidal action of many β-lactam antibiotics and makes the organism resistant to these molecules. For this reason, the β-lactamase inhibitors have long been regarded as promising targets from a medicinal viewpoint. A comparison between the kinetic characteristics of β-lactamase and penicillin-sensitive enzymes (carboxypeptidase and transpeptidase) is of interest in this respect. β-Lactamases very efficiently hydrolyze β-lactam in contrast to penicillin-sensitive enzymes [high k_4 in Eq. (9)].

It would be valuable to develop compounds affording a stable acyl-β-lactamase. Clavulanic acid (*40*) is a natural product discovered in a *Streptomyces* strain and acts as a specific inhibitor of β-lactamase. Fisher and Knowles indicated the possibility of the formation of a long-living acyl-enzyme in the catalytic pathway of β-lactamase having a serine residue in the active site. The inhibition of β-lactamase by clavulanic acid is suggested to be a consequence of the subsequent formation of *41*[61].

The synthetic penicillin sulfone [sulbactam, (*42*)] has been shown to act as a β-lactamase inhibitor. The inhibition is based on a similar mechanism as proposed above [62]. A prodrug, sultamicillin (*43*), which combines sulbactam and amoxacillin by labile

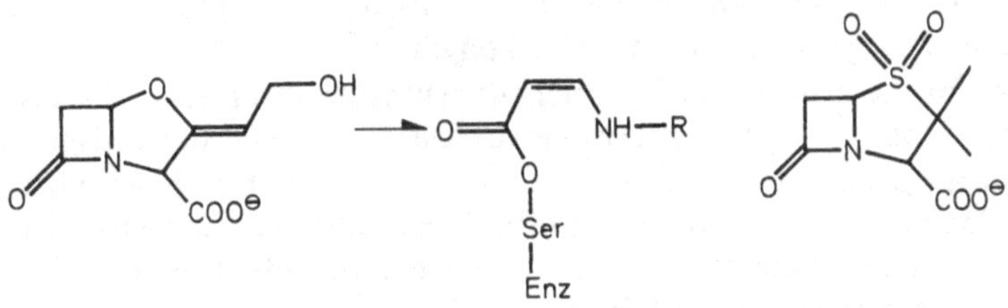

40 *41* *42*

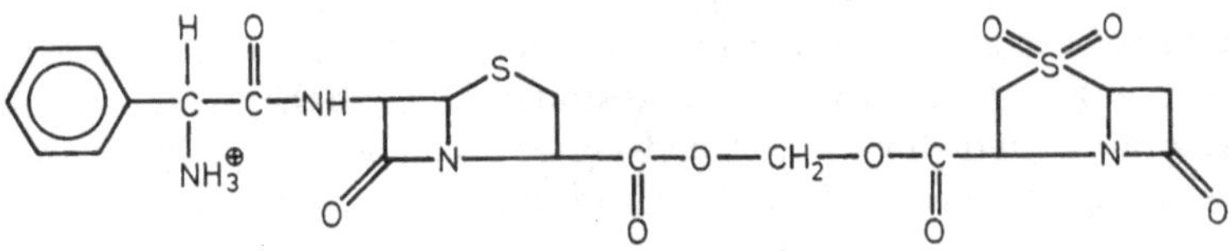

43

linkage is designed to deliver both a lactam and a β-lactamase inhibitor [63]. A possible mechanism-based inhibitor for β-lactamase *44* (Fig. 2) was proposed [64]. It is suggested that the inhibition is initiated by the formation of the acyl enzyme *45* in which the concomitant loss of a fluoride ion is taking place as shown in Fig. 2.

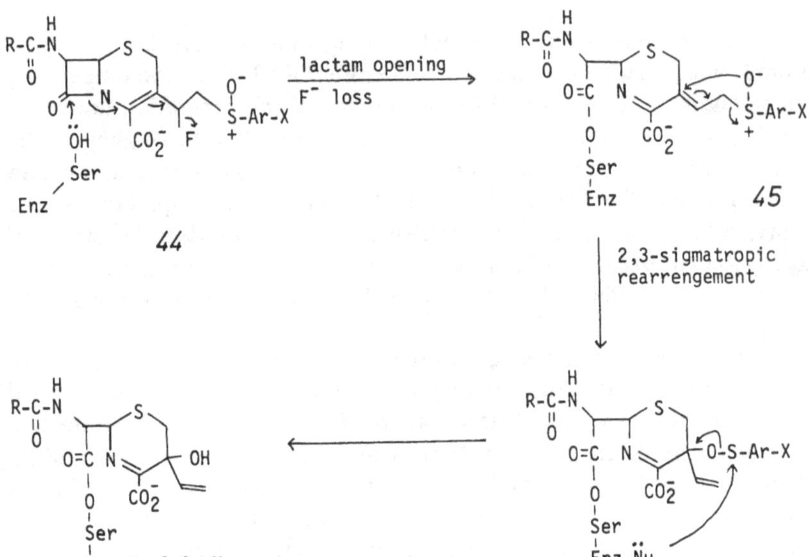

Fig. 2. Proposed mechanism for the inhibition of β-lactamase by *44*

5 Design of Trypsin Substrates of a New Type — Inverse Substrates

Enzymes which catalyze the hydrolysis of the unit linkage of sequential residues of oligomers or polymers determine their substrate specificity by recognizing the particular unit residue in the sequential chain as well as the direction of the chain. For example, ribonuclease cleaves the 3′-phosphate of a pyrimidine nucleotide residue but not the 5′-phosphate, and trypsin hydrolyzes peptide bonds which involve the arginine or lysine residue at the carbonyl end but not at the amino end. This is also the case for the hydrolysis of a variety of synthetic substrates and quasi-substrates (Sect. 4.1). Synthetic trypsin substrates are ester or amide derivatives in which the site-specific group (positive charge) is contained in their carbonyl portion.

Compounds which violate this empirical rule have not been observed, though some attempts have been made to design such compounds. Therefore, it has been generally considered that any modification in the fundamental architecture of the substrate molecule would cause a loss of susceptibility.

In our early work, esters of *p*-amidinobenzoic acid (*8*) were shown to behave as specific substrates of trypsin, as mentioned in Sect. 4.1. Esters *8* have the same molecular arrangement as normal-type substrates although they have a simplified structure lacking an asymmetric carbon and an α-acylamide group. In an extension of this investigation, we designed esters of an inverted structure, namely acyl derivatives of *p*-amidinophenol (*46*).

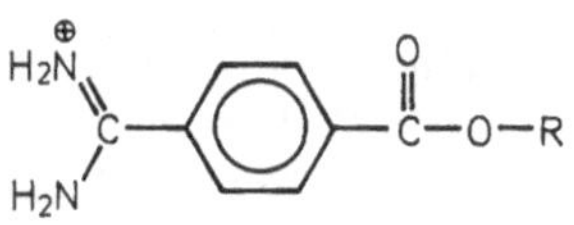

8

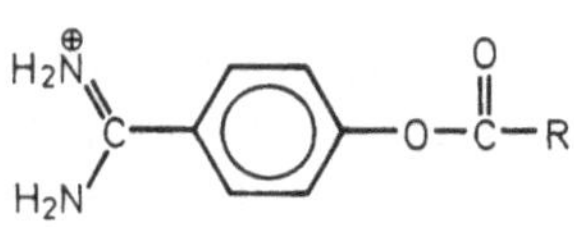

46

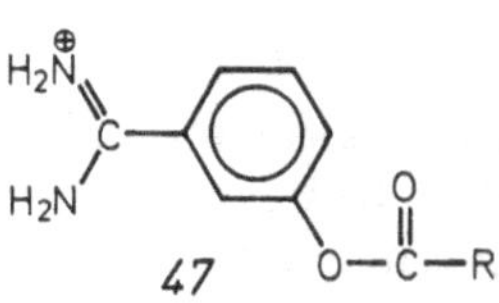

47

It was found that *46* behaves as an exceptional substrate of trypsin, showing a the reaction mode which had not been observed before. Fig. 3 shows the time course of the tryptic catalysis of *46* monitored by the amidinophenol liberation under the condition that the substrate is in much higher concentration than the enzyme. After rapid mixing of enzyme and substrate, a rapid acylation step is observed and a slow deacylation then follows. The kinetics follow a Michaelis-Menten equation: strong binding affinity, efficient acylation, and rate-determining slow deacylation steps, which are exactly the same as those of normal-type substrates. As a result, the accumulation of the acyl enzyme intermediate (*EA*) is realized in the course of the steady-state hydrolysis [cf. Eq. (6)].

For the esters *46*, the site-specific group for the enzyme, charged amidinium, is not included in the acyl moiety but in the leaving portion, and so these esters were termed "inverse" substrates with respect to their kinetic parameters [65]. Kinetic parameters for some inverse substrates are listed in Table 3 together with those for a normal type (8) and the *meta*-isomer *47*. *p*-Amidinophenyl acetate, for example, exhibits a binding constant of 10^{-5} M, an efficient acylation stage with a rate constant of $17\ s^{-1}$, and a slow deacylation. In contrast, the *meta*-isomers were found to be very poor substrates, probably because of unfavorable positioning of the carbonyl, which is shown by the small acylation rate constant. The normal-type substrate has a binding

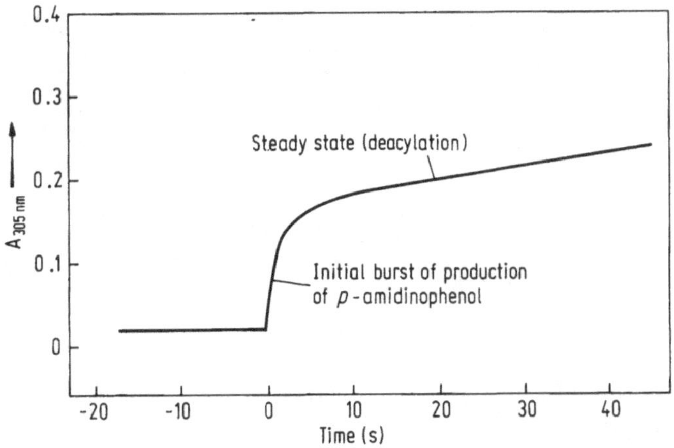

Fig. 3. Time course of the trypsin-catalyzed hydrolysis of *p*-amidinophenyl acetate (*46*; R = CH_3) at pH 8.0, 25 °C. Concentrations of enzyme and ester are 10 μM and 0.7 mM, respectively

Table 3. Kinetic parameters for tryptic hydrolysis of "inverse" and normal type substrates[a]

Compound	K_s (M × 10^5)	k_2 (s^{-1})	k_3 (k_{cat})[b] (s^{-1} × 10^4)	k_{spont} (s^{-1} × 10^6)
46; R = CH_3	3.87	17.0	92.6	26.0
47; R = CH_3	3.03	0.03	(49.8)	15.8
8	0.503	30.4	653	217
AcONP[c]	2100	1.5	130	nd[b]

[a] Reaction was carried out in 0.05 M tris buffer containing 0.02 M $CaCl_2$ at pH 8.0, 25 °C;
[b] Overall k_{cat} (k_3 is not much smaller than k_2);
[c] *p*-Nitrophenyl acetate;
[d] Not determined

constant and acylation rate constant comparable to those of inverse substrates. Non-specific *p*-nitrophenyl acetate exhibits very poor binding and insufficient acylation.

The reaction process of trypsin-catalyzed hydrolysis of the inverse substrates is illustrated in Fig. 4. Here the process is compared to that of normal-type substrates. After specific binding and efficient acylation, the site-specific amidinophenyl moiety is cleaved (leaving group) to give the acyl enzyme in a very specific manner. As a result, inverse substrates are expected to be applicable as a general method for "specific" introduction of any acyl group of "non-specific structure" into the trypsin active site.

For the first time, inverse substrates provide a general method for the specific introduction of an acyl group into the trypsin active site without recourse to cation-containing acyl compounds. The preparation of various new acyl enzymes is expected to lead to the discovery of novel features of the enzymatic reaction mechanism. In addition, any desired reporter groups might be specifically introduced into the trypsin

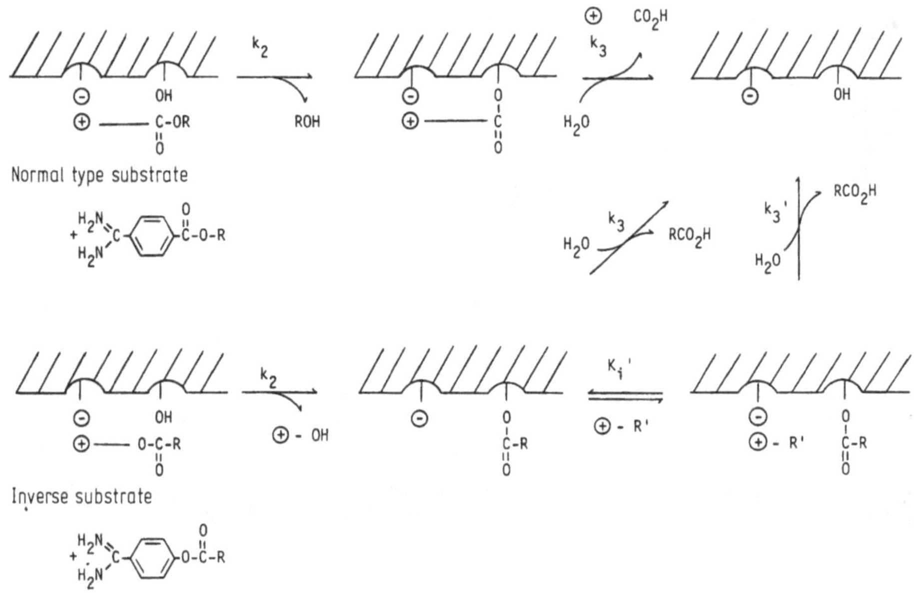

Fig. 4. Reaction sequences of trypsin with normal type and "inverse" substrates. The hydroxyl function and negative charge represent the catalytic residue (Ser-195) and the binding residue (Asp-189) at the active site, respectively. The acyl trypsin-ligand complex (low right) formed in the presence of a cationic compound

active site and these acyl enzymes will provide information on the structure of the active site vicinity.

It is of special value to extend the "inverse" concept further to trypsin-like enzymes. Inverse substrates of these biologically important enzymes could be candidates for clinically useful substances. In the following sections, various aspects of the applicabilities are briefly described.

5.1 Applications to the Studies on Structure and Function of Trypsin

Enzymatic kinetics for the esters derived from acetylamino acids and acetylpeptides were studied. Even in the case of D-amino acid derivatives, catalysis was found to give acyl enzyme intermediates in a very specific manner, as shown in Table 4. Inverse substrates are artificial substances affording acyl enzymes, and accordingly their K_s and k_2 values have no significant meaning in a physiological sense. Hydrolytic processes of these aminoacyl or peptidyl enzyme intermediates (k_3), however, are the stages of enzymatic action itself, because trypsin is a proteolytic enzyme leading to an aminoacyl and peptidyl enzyme as an intermediate. These parameters for D-amino acid derivatives which never have been analyzed by conventional substrates because of their unsusceptibility at the acylation step, have now been determined and the resulting values have been evaluated [66,67].

An additional characteristic of inverse substrates is also shown in Fig. 4. The acyl enzyme formed from the inverse substrate lacks a site-specific cationic residue with

Table 4. Kinetic parameters for the trypsin-catalyzed hydrolysis of inverse substrates at pH 8.0, 25 °C

Compound	K_s (M × 10^5)	k_2 (s^{-1})	k_3 (s^{-1})
46			
R = Ac-Gly	2.9	19	0.61
R = Ac-$(Gly)_2$	2.7	21	1.3
R = Ac-$(Gly)_3$	4.7	15	1.5
R = Ac-L-Ala	4.6	4.8	2.1
R = Ac-D-Ala	3.0	7.0	0.012
R = Ac-L-Ala-Gly	3.6	9.6	1.8
R = Ac-D-Ala-Gly	1.9	6.6	0.28

which the binding site interacts. Therefore, the acyl enzyme is capable of accepting an external, charged molecule at this vacant binding site to form an acyl enzyme-charged molecule complex if sufficient concentration of cationic ligand is added and if this vacant cavity is large enough to allow coexistance of both the acyl residue and the charged molecule. The dissociation constant of this complex was denoted as K_i' as shown in the figure. The deacylation rate constant for this complex, k_3', is expected to be different from that of the simple acyl enzyme without the ligand. The presence of the cationic compound caused a rate acceleration in the overall catalytic rate of inverse substrates. This acceleration occurred at the rate-determining deacylation stage. Cationic compounds which are known as competitive inhibitors for trypsin generally exhibit a rate acceleration effect and the effect depends on the ligand concentration [68,69]. Analysis of the rate enhancements observed with a variety of inverse substrates and cationic ligands refined the mechanistic understanding of the catalytic efficiency of trypsin [70].

p-Amidinophenyl esters carrying a fluorophore [71], an optically active chromophore [72], or a stable, free radical [73] have been synthesized. All of these esters exhibited a strong binding affinity and an efficient acylation step. Isolation of acyl trypsin was successfully carried out by the general procedure as follows: About 20 equivalents of a substrate were mixed with the enzyme at room temperature at pH 8.0. After standing for several minutes, the pH was dropped by addition of diluted hydrochloric acid to around 2. The reaction mixture was gel-filtered and subsequently lyophilized.

The microenvironment of the trypsin active site was estimated by spectrometric analysis of these acyl trypsin preparations.

5.2 Inverse Substrates for Trypsin-Like Enzymes — Medicinal Applicabilities

It is well known that the specificity of an enzyme such as thrombin and plasmin is very close to that of trypsin. In this respect, inverse substrates for trypsin also are expected to be susceptible to the catalysis by these enzymes. In the kinetic analysis of trypsin-like enzymes toward *p*-amidinophenyl esters, it was found that the "inverse" concept is also applicable to thrombin, plasmin, urokinase, kallikrein, and trypsins from various origins [74–75]. These enzymes are not distinctively different from bovine-trypsin in their binding constants and acylation rate constants. Deacylation rate constants, however, are more variant. Trypsin-like enzymes are known to have a key role in such important physiological phenomena like coagulation and fibrinolysis and therefore the compounds which are capable of discriminating between these enzymes could be of therapeutic value.

The active site structure of trypsin-like enzymes is considered to be very similar to that of bovine trypsin, yet little is known about them. Refinement of these structures is important also for the purpose of designing physiologically active substances. With a view to comparing the spatial requirements of active sites of these enzymes, dissociation constants of the acyl enzyme-ligand complex, K_i', which were defined before, were successfully analyzed [illegible]. By taking advantage of inverse substrates which have an unlimited choice of the acyl component, development of stable acyl enzymes could be possible. These transient inhibitors for trypsin-like enzymes could be candidates for drugs. In this respect, the determination of the deacylation rate constants for the plasmin- and thrombin-catalyzed hydrolyses of various esters were undertaken [77].

A new approach to thrombosis therapy using acyl plasmins has been reported by Smith et al. [78]. Acyl plasmin is catalytically inert and unable to react with plasma inhibitors but still can bind to a fibrin clot. Thus, after the administration, acyl plasmin can circulate without being trapped by the inhibitors and can come into contact with fibrin. Deacylation may then occur to give a fibrin-plasmin complex and this active enzyme is expected to lead to fibrinolysis. The preparation of acyl plasmin of appropriate stability was realized by using the general procedure for the specific synthesis of an acyl enzyme — the "inverse substrate" method.

5.3 Considerations on the Concept of Inverse Substrates

Among a number of experimental results in which the kinetic behavior of protelytic enzymes toward a variety of synthetic substrates and inhibitors have been tested, some seemingly irrational enzymatic responses were observed. Example of these responses will be discussed from the viewpoint of the imperfectness of the enzymatic recognition. The existence of inverse substrates might be due to such an imperfectness or allowance in the recognition rigidity of the enzyme.

So-called "non-productive" or "wrong way" binding must be the binding mode of physiological meaning in which an enzyme prevents wrong substrates from being involved in catalysis. A typical example of the binding is shown in the interaction of such a protease, as e.g., chymotrypsin with D-amino acid derivatives. However, "non-productive binding" formed between chymotrypsin and its substrate, acetyltyrosine-anilide, is somewhat different [79]. As is known, chymotrypsin exhibits its substrate specificity toward aromatic amino acids, and in this instance chymotrypsin cannot discriminate between either the aromatic residue of a tyrosine side chain or an anilide moiety even if the substrate is L-configurated (Fig. 5e).

The binding constant of the substrate acetyl-L-leucyl-L-tyrosine methylamide to pepsin (K_m) is reported as 2.7 mM and the binding of the inhibitor acetyl-D-tyrosyl-D-leucine methylamide (K_i) as 5.8 mM. The binding shown in Fig. 6 was proposed for the reason that both binding constants are almost identical. This assumption is based upon the idea that the space-filling structure of leucyltyrosine in the L-configuration is similar to that of the reversed sequence, tyrosylleucine, in the D-configuration. A

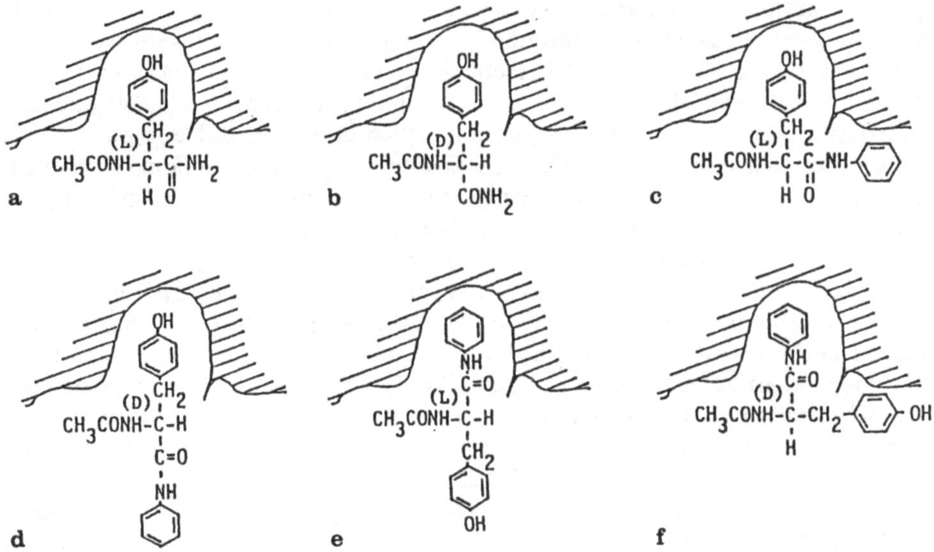

Fig. 5. Productive [**a**), **c**)] and non-productive [**b**), **d**)–**f**)] binding of tyrosine derivatives to chymotrypsin. Catalytic residues in the active site are illustrated as a sharp edge

term "retro-enantiomer" was proposed for this concept [80]. This might be another example in which the enzyme has been misled. The peptide bond in this case (Fig. 6a) is resistant to hydrolysis because its orientation to the catalytic residue of the enzyme is not properly attained.

Attempts for designing inverse-type compounds have been reported by other research groups. In case of chymotrypsin, Jones et al. [81] prepared certain esters with alcohol components which imitate tryptophan and phenylalanine residues. Attempts have also been made by Hartman et al. [82] and Muramatsu et al. [83]. They prepared aminobutanol acetate for trypsin. But all these compounds were found not to be hydrolyzed appreciably under the chosen conditions. Although the enzyme function is not always

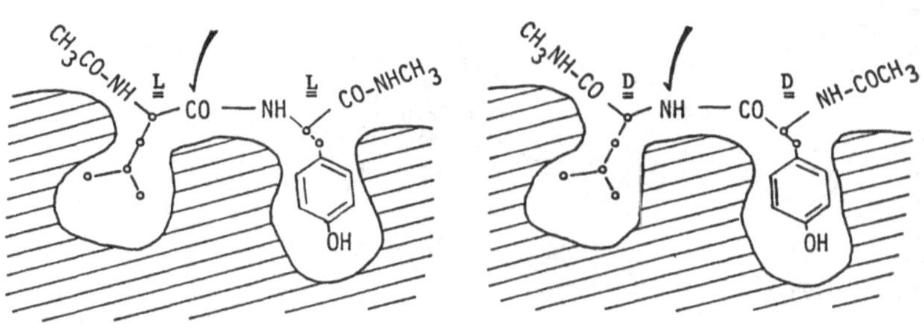

a Ac-L-Leu-L-Tyr-$NHCH_3$ b Ac-D-Tyr-D-Leu-$NHCH_3$

Fig. 6. Binding of specific substrate **a**) and its "retro-enantiomer" **b**) to pepsin. Arrow represents the proximity of the catalytic residues

perfect, enzymes are still able to discriminate such derivatives. It can be assumed, therefore, that in designing extraordinary compounds, such as inverse substrates, some adjustments in the chemical reactivity are needed. Our leaving group, *p*-amidinophenol, is chemically different from that of the above authors (phenol vs. aliphatic hydroxyl group). Phenol esters are generally much more susceptible to nucleophilic substitution than esters of aliphatic alcohols. Furthermore, the *p*-amidino substituent has an electron withdrawing character nearly equal to the *p*-nitro group [84]. In our case, *p*-amidinophenyl esters might satisfy both conditions: the spatial requirements of the active site of the enzyme and the chemical reactivity itself.

The involvement of several residues to serve as general acid and base is well recognized in the catalytic processes of trypsin and chymotrypsin. In the acylation stage, inverse substrates will be distinguished from normal-type substrates by the assistance of these residues. A comparison is made in Fig. 7. In the case of normal-type substrates, participation of the general acid on the leaving group (OR) will assist the formation of the acyl enzyme. In contrast, for inverse substrates the leaving group does not come into contact with the general acid residue, because OR is oriented in the opposite direction. It is concluded therefore that our leaving group chemically compensates for the inherent disadvantage of the enzymatic process with the inverse-type substrates. The reason why compounds reported by Jones et al., Hartman et al., and Muramatsu et al. behaved simply as competitive inhibitor will thus be explained.

It is perhaps worth reconsidering the status of conventional substrates of chymotrypsin derived from *p*-nitrophenol in terms of the "inverse" concept. *p*-Nitrophenyl acetate, a well-known substrate for chymotrypsin, is an active ester with an aromatic moiety (specific residue for the enzyme) in its leaving portion. This could be considered as a sort of inverse substrates, though its binding affinity is not excellent. In this respect, 2-hydroxy-5-nitro-α-toluenesulfonic acid sulfone could be considered as a hybrid of normal and inverse ones [85]. The reaction process of inverse substrates is essentially the same as that of conventional ester substrates following the whole enzymatic process to regenerate the original enzyme. This is a characteristic feature of inverse substrates which is not satisfied by mechanism-based and affinity labeling inhibitors. The most striking characteristic is to afford an acyl enzyme without re-

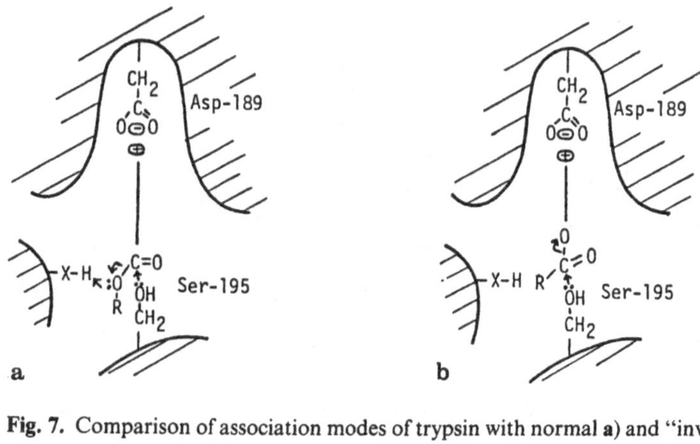

Fig. 7. Comparison of association modes of trypsin with normal **a)** and "inverse" **b)** substrates

course to the structure of the acyl moiety. As a result, any desired acyl groups might be introduced specifically into the enzyme active site.

Further search for inverse substrates other than *p*-amidinophenyl esters has been carried out and it has been found that esters derived from *p*-aminomethylphenol and *p*-guanidinophenol were also eligible as a substrate of trypsin and trypsin-like enzymes [75,86]. We have also found that trimethylaminobutanoic acid *p*-nitrophenyl ester is an inverse substrate for butyrylcholinesterase [87,88]. Application of the inverse concept to thiol enzymes was also successful: *p*-amidinophenyl esters were found to be substrates for clostripain [74], a thiol enzyme with trypsin-like specificity. Although the design of inverse-type substrates seems not always possible for a variety of hydrolytic enzymes, this new concept could provide potential means for certain enzymes to both: fundamental study and application.

6 Conclusion

The design of enzyme-specific compounds is one of the most promising subjects of our time. It is not only significant in the elucidation of structure-function relationship of enzymes but also useful as a methodology of drug design. Although the number of drugs developed so far through the methodology is not large, the subject will become increasing significant for the purpose. It has been generally assumed that enzymes exhibit substrate recognition in a very strict manner, but we may fortunately conclude that enzymatic recognition is not completely perfect in some cases. Consequently, the design of compounds which trick enzymes is possible, and mimics like "inverse substrates" have been found. Distinct from simple competitive inhibitors, these mimics which interact with enzymes in a sophisticated manner will provide new concepts for the design of clinically useful substances. The rational approach for drug design will grow parallel with our knowledge of various *in vivo* factors as well as with the development of new concepts for drug design.

7 References

1. Rozenblit, A. B.: Computer-assisted drug design. Strategy and alogrithms, in: Strategy in Drug Research, (ed.) Buisman, J. A. K., Vol. 4, p. 287, Amsterdam, Elsvier 1982
2. Wilchek, M. et al.: Methods in Enzymol. *104*, 3 (1984)
3. Horejsi, V.: ibid. *104*, 275 (1984)
4. Bayley, H., Knowles, J. R.: ibid. *46*, 69 (1977)
5. Peach, M. J.: Physiol. Rev. *57*, 313 (1977)
6. Cushman, D. W. et al.: Biochemistry *16*, 5484 (1977)
7. Cushman, D. W. et al.: Angiotensin-converting enzyme inhibitors, in: Enzyme Inhibitors as Drugs, (ed.) Sandler, M., p. 231, London, Macmillan Press 1980
8. Powell, J. R. et al.: Abstr. Eur. Meet. Hypertens. 1st. Milan 355 (1983)
9. Unger, T. et al.: Eur. J. Pharmacol. *78*, 411 (1982)
10. McEvoy, F. J. et al.: J. Med. Chem. *26*, 381 (1983)
11. Thorsett, E. D. et al.: Biochem. Biophys. Res. Comm. *111*, 166 (1983)
12. Abiko, Y.: Medicinal Pharmacy *10*, 8 (1976)
13. Burchall, J. J.: Mechanism of action of antimicrobial and antitumor agents, in: Antibitics, (eds.) Corcoran, J. W., Hahn, F. E., Vol. III, p. 304, Berlin, Springer 1975
14. Beddell, C. R.: Chem. Soc. Rev. *13*, 279 (1984)

15. Kuyper, L. K. et al.: J. Med. Chem. *28*, 303 (1985)
16. Blaney, J. M. et al.: Chem. Rev. *84*, 333 (1984)
17. Edy, J. et al.: Thromb. Res. *8*, 513 (1977)
18. Kanaoka, Y. et al.: Chem. Pharm. Bull. *33*, 1721 (1985)
19. Kanaoka, Y.: Angew. Chem., Int. Ed. Engl. *16*, 137 (1977)
20. Jameson, G. W. et al.: Biochem. J. *131*, 107 (1973)
21. Chase, T. Jr., Shaw, E.: Methods in Enzymol. *19*, 20 (1975)
22. Coleman, P. L. et al.: ibid. *45*, 12 (1976)
23. Tanizawa, K. et al.: Biochem. Biophys. Res. Comm. *32*, 893 (1968)
24. Chase, T. Jr., Shaw, E.: ibid. *29*, 508 (1967)
25. Livingston, D. C. et al.: Biochemistry *20*, 4298 (1981)
26. Pauling, L.: Chem. Eng. News *24*, 1375 (1946)
27. Lienhard, G. E.: Science *180*, 149 (1973)
28. Wolfenden, R.: Methods in Enzymol. *46*, 15 (1977)
29. Matthews, D. A. et al.: J. Biol. Chem. *250*, 7120 (1975)
30. Lindquist, R. N., Nguyen, A. C.: J. Am. Chem. Soc. *99*, 6437 (1977)
31. Umezawa, H. et al.: J. Antibiotics *25*, 267 (1972)
32. Aoyagi, T. et al.: ibid. *22*, 283 (1969)
33. Okura, A. et al.: ibid. *28*, 337 (1975)
34. Thompson, R.: Methods in Enzymol. *46*, 220 (1977)
35. Brayer, G. D. et al.: Proc. Natl. Acad. Sci. (USA) *76*, 96 (1979)
36. Kuramochi, H. et al.: J. Biochem. *86*, 1403 (1979)
37. Brodbeck, U. et al.: Biochim. Biophys. Acta *567*, 357 (1979)
38. Geratz, J. D.: Arch. Biochem. Biophys. *118*, 90 (1967)
39. Tanizawa, K. et al.: Z. Physiol. Chem. *366*, 87 (1985)
40. Blake, C. C. F. et al.: Proc. Roy. Soc. B167, 378 (1967)
41. Secemski, I. I., Lienhard, G. E.: J. Biol. Chem. *249*, 2932 (1974)
42. Shaw, E.: Methods in Enzymol. *25*, 387, 655 (1972)
43. Kettner, C., Shaw, E.: ibid. *80*, 826 (1981)
44. Hess, G. M. et al.: Biochemistry *11*, 3787 (1972)
45. Zispel, N. et al.: Biocem. Biophys. Res. Comm. *58*, 457 (1974)
46. Plummer, T. H. Jr. et al.: J. Biol. Chem. *244*, 5246 (1969)
47. Lawson, W. B., Shramm, H.-J.: Biochemistry, *4*, 377 (1965)
48. Walsh, C. T.: Ann. Rev. Biochem. *53*, 493 (1984)
49. John, R. A.: Enzyme-Induced Inactivation of Pyridoxal Phosphate-Dependent Enzymes: Approaches to the design of specific inhibitors, in: Enzyme Inhibitors as Drugs, (ed.) Sandler, M., p. 73, London, Macmillan Press 1980
50. Fowler, L. J. et al.: Biochem. J. *130*, 569 (1972)
51. Wolfenden, R.: Methods in Enzymol. *46*, 15 (1977)
52. White, E. H. et al.: J. Am. Chem. Soc. *97*, 2290 (1975)
53. Westkamper, R., Abels, R.: Biochemistry *22*, 2356 (1983)
54. Daniels, S. et al.: J. Biol. Chem. *258*, 15046 (1983)
55. Wilson, I. B. et al.: J. Biol. Chem. *236*, 1498 (1961)
56. Moorman, A., Abeles, R.: J. Am. Chem. Soc. *104*, 6785 (1982)
57. Chase, T. Jr., Shaw, E.: Biochemistry *8*, 2212 (1969)
58. Tipper, D. J., Strminger, J. L.: Proc. Natl. Acad. Sci. (USA) *54*, 1133 (1965)
59. Frére, J. E. et al.: Eur. J. Biochem. *57*, 343 (1975)
60. Ishii, S., Kasai, K.: Methods in Enzymol. *80*, 842 (1981)
61. Fisher, J. F., et al.: Biochemistry *17*, 2180 (1978)
62. Fisher, J. F., Knowles, J. R.: The Inactivation of β-Lactamase by Mechanism-Based-Reagents, in: Enzyme Inhibitors as Drugs, (ed.) Sandler, M., P. 209, London, Macmillan Press 1980
63. Hartley, S., Wise, R.: J. Antimicrob. Chemother. *10*, 49 (1982)
64. Firestone, R. A.: US Patent 4342758 (1982)
65. Tanizawa, K. et al.: J. Am. Chem. Soc. *99*, 4485 (1977)
66. Fujioka, T. et al.: Chem. Pharm. Bull. *28*, 1899 (1980)
67. idem.: J. Biochem. *89*, 637 (1980)
68. Tanizawa, K., Kanaoka, Y.: Experientia *35*, 16 (1979)
69. Tanizawa, K. et al.: J. Biochem. *8*, 417 (1980)

70. Tanizawa, K. et al.: ibid. *92*, 945 (1982)
71. Nakayama, H. et al.: J. Am. Chem. Soc. *102*, 3214 (1980)
72. Nakayama, H. et al.: Eur. J. Biochem. *112*, 403 (1980)
73. Fujioka, T. et al.: Biochim. Biophys. Acta *612*, 205 (1980)
74. Nozawa, M. et al.: J. Pharm. Dyn. *4*, 559 (1981)
75. idem.: ibid. *3*, 213 (1980)
76. idem.: J. Biochem. *91*, 1837 (1982)
77. McLaren, A. B., Tanizawa, K.: Aust. J. Biol. Sci. *37*, 205 (1984)
78. Smith, R. et al.: Nature *290*, 505 (1981)
79. Fastrez, J., Fersht, A. R.: Biochemistry *12*, 1067 (1973)
80. Shemyakin, M. M. et al.: Angew. Chem., Int. Ed. Engl. *8*, 492 (1969)
81. Jones, J. B. et al.: Biochim. Biophys. Acta *341*, 284 (1974)
82. Hartman, H., Holler, E.: Eur. J. Biochem. *16*, 80 (1970)
83. Muramatsu, M. et al.: J. Biochem. *62*, 408 (1967)
84. Wang, C., Shaw, E.: Arch. Biochem. Biophys. *150*, 259 (1972)
85. Haidema, J. H., Kaiser, E. T.: J. Am. Chem. Soc. *90*, 1860 (1968)
86. Nakano, M. et al.: Chem. Pharm. Bull. *28*, 2212 (1980)
87. Nozawa, M. et al.: Biochim. Biophys. Acta *611*, 314 (1980)
88. idem.: J. Pharm. Dyn. *3*, 321 (1980)

Recent Developments in the Field of Biologically Active Peptides[1]

Gunter Schmidt

EP-FE Chemisch wissenschaftliches Labor, BAYER AG
D-5600 Wuppertal 1, FRG

Table of Contents

[1] *Dedicated to Prof. Helmut Zahn on the occasion of his 70th birthday.*

The abbreviations and symbols for amino acids and peptides used in this article comply with the rules published by the IUPAC Commission for Biochemical Nomenclature; see: J. Biol. Chem. *245*, 6489 (1970); ibid. *247*, 977 (1972); ibid. *250*, 3215 (1975); Biochem. J. *126*, 773 (1972). The amino acids are in the L-configuration; in the case of D-amino acids the abbreviations are prefixed by the letter D.

1 Introduction — Methods of Peptide Synthesis

Peptide synthesis has once again stepped into the limelight in the last 10 years with the discovery of neuropeptides, peptides of the renin-angiotensin system, and immunoactive peptides. Above all, the solid-phase peptide synthesis of R. B. Merrifield, besides conventional peptide syntheses, has made an inestimable contribution to the production of biological and medically important polypeptides.

Although peptides still have little market potential within the sector of pharmaceutical chemistry, one can expect their use for therapeutic purposes to increase in the future in view of their importance as bioregulators.

Synthetic peptides such as oxytocin, vasopressin, ACTH, calcitonin, secretin, somatostatin, cyclosporine, and insulin are already in clinical use.

One major disadvantage of peptide active substances is their denaturation and enzymatic degradation in the gastrointestinal tract, which mean that at present only parenteral, sublingual, or intranasal administration is possible.

The present article gives an overview of the chemistry, biochemistry, and physiology of interesting natural and synthetic peptides.

L-amino acids, and in many cases also non-proteinogenic amino acids such as D-amino acids, α-aminoisobutyric acid, isovaline, β-alanine, and N-methylamino acids, serve as the raw materials for the production of peptides. The formation of the peptide linkage between the amino acids takes place in the following ways:

1. Conventional method: stepwise synthesis or fragment condensation using an optimized choice of protecting group combination and the most favorable coupling methods [1-8],

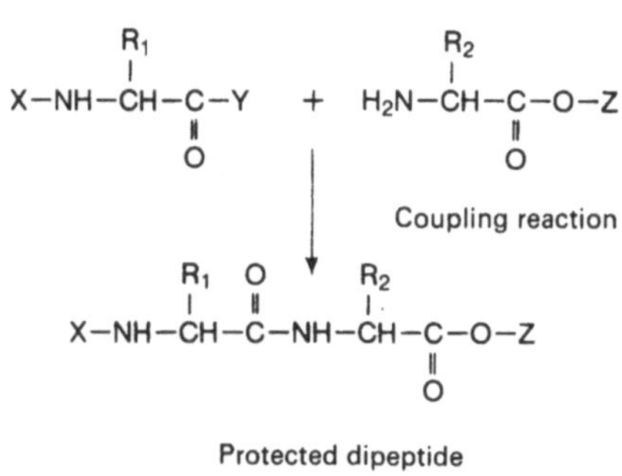

Fig. 1. Principle of peptide synthesis

X: amino-protecting group (e.g. *t*-butyloxycarbonyl);
Y: activating group (e.g. $-N_3$, $-C_6H_4$-p-NO_2,

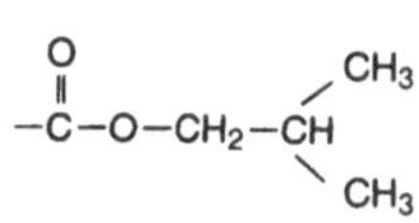

Z: carboxyl-protective group (e.g. $-CH_3$, $-CH_2$ 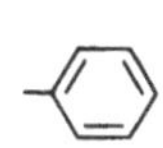).

2. Solid-phase method of Merrifield [9-11] and the liquid-phase method of Mutter [12],
3. Protease-catalyzed peptide syntheses [13-16],
4. DNA recombination, i.e. bacterial production of peptide hormones, e.g. insulin, somatostatin, or GRF (growth-hormone-releasing factor) [17,18].

In recent times peptide analogs have also been produced by a modification of the CO—NH bond (peptide backbone modification) [19]:

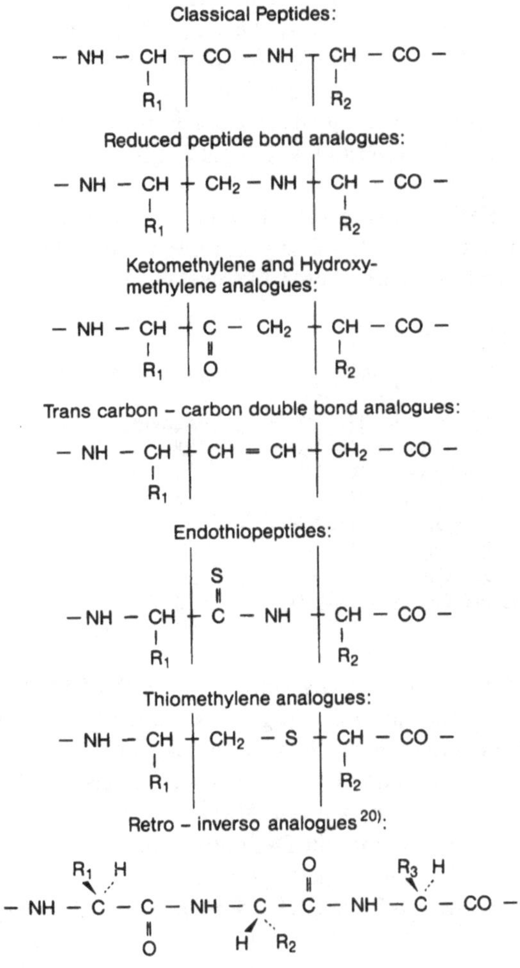

According to the mode of action of the peptide and the pathway from the hormone-active cells to the target organ, the following peptide groups can be distinguished [5,21].

2 Neuropeptides

The biochemistry, physiology, pharmacology, and synthesis of the neuropeptides (peptides of the central nervous system) have been in the mainstream of research on vegetative and hormonal regulation in man and in animals in the last 10 years [22].

40 years elapsed between the discovery of subtance P by Euler and Gaddum in 1931 and its synthesis. Many neuropeptides have been found in nervous system in the last 12 years.

2.1 Substance P (SP)

SP *1* was discovered in the brain and the intestinal tract of man, mammals, and birds, and was synthesized in 1971 by Tregear [23] by the solid-phase method.

It has the effects typical of the kinins: e.g. stimulation of the smooth muscle and lowering of the blood pressure due to vasodilation.

SP, which can function as a neurotransmitter in various brain regions, suppresses the action of morphine and endorphins and is thought to play a protective role against stress-determined disturbances.

The structure-activity relationships in the SP molecule show that the C-terminal pentapeptide represents the active center. The efficacy is increased by stepwise chain prolongation of the C-terminal pentapeptide (Table 1).

Table 1. Substance-P derivatives: relative activities on guinea-pig ileums (GPI)

1 2 3 4 5 6 7 8 9 10 11	Activity	
Arg-Pro-Lys-Pro-Gln-Gln-Phe-Phe-Gly-Leu-Met-NH_2	100	(1)
Pro-Lys-Pro-Gln-Gln-Phe-Phe-Gly-Leu-Met-NH_2	60	(2)
Lys-Pro-Gln-Gln-Phe-Phe-Gly-Leu-Met-NH_2	160	(3)
Pro-Gln-Gln-Phe-Phe-Gly-Leu-Met-NH_2	200	(4)
Gln-Gln-Phe-Phe-Gly-Leu-Met-NH_2	125	(5)
Gln-Phe-Phe-Gly-Leu-Met-NH_2	100	(6)
Phe-Phe-Gly-Leu-Met-NH_2	2	(7)

The myotropic effect (stimulation of the smooth muscle) of SP fragments *3*, *4*, and *5* is greater than that of substance P. Fragments *6* already exerts the complete biological activity. The peak activity is reached with fragment *4*.

The SP derivative pyroGlu-Phe-Phe-Gly-Leu-Met-NH_2 is one of the most active compounds in the vasodepressor-response test [24].

Some substance P derivatives that contain D-amino acids, e.g. [Arg^6, D-Trp^{10}]-SP (6—11) and [D-Pro^4, D-$Trp^{7,9}$]-SP (4–11) act as strongly competitive antagonists [25].

The retro-inverso SP derivative *8* [$pyroGlu^6$, gPh^8, $mGly^9$]-SP (6–11) (g = gem.

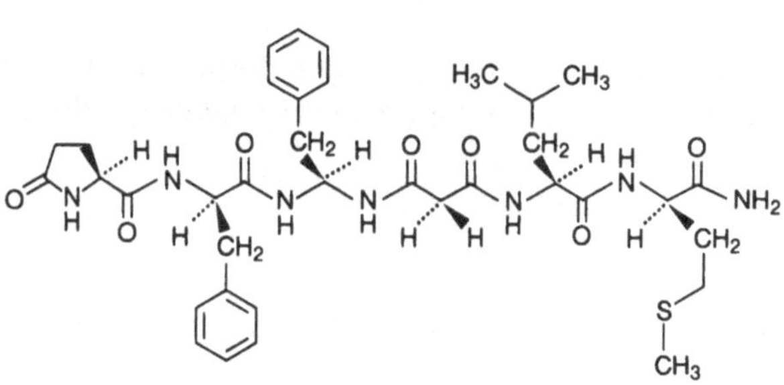

(8)

diamino residue, *m* = malonic acid residue) is a total agonist of substance P [26] and is stable to proteolytic cleavage [27].

2.2 Neurokinins

Neurokinins *9* and *10*, which were isolated from porcine spinal cord extracts and synthesized in 1984 by Munekata et al. [28], show a strong hypotensive effect like substance P.

H-His-Lys-Thr-Asp-Ser-Phe-Val-Gly-Leu-Met-NH_2 (9)

H-Asp-Met-His-Asp-Phe-Phe-Val-Gly-Leu-Met-NH_2 (10)

2.3 Neurotensin (NT)

pyroGlu-Leu-Tyr-Glu-Asn-Lys-Pro-Arg-Arg-Pro-Tyr-Ile-Leu (11)

Neurotensin *11*, which was isolated from bovine small intestine by Carraway and Leeman [29] in 1973, causes, in addition to the typical plasma kinin effects (lowering of the blood pressure, contracting action on the intestine and uterus), an increase in the LH and FSH secretion without influencing the release of somatotropin or thyrotropin.

St.-Pierre et al. [30] synthesized many NT fragments that are biologically active in the cardiovascular system. NT 8–13 shows the complete range of action of the native NT.

2.4 Endorphins, Enkephalins, Dynorphin, and Dermorphin (Opioid Peptides)

The first endogenous peptides *16* and *17* with morphine-like activity were isolated from human and animal nerve tissue by Hughes and Kosterlitz [31] in 1975.

Tyr-Gly-Gly-Phe-Met-Thr-Ser-Glu-Lys-Ser-Gln-Thr-Pro-Leu-Val-Thr (12)

Tyr-Gly-Gly-Phe-Met-Thr-Ser-Glu-Lys-Ser-Gln-Thr-Pro-Leu-Val-Thr-Leu-Phe-Lys-Asn-Ala-Ile-Ile-Lys-Asn-Ala-Tyr-Lys-Lys-Gly-Glu (13)

Tyr-Gly-Gly-Phe-Met-Thr-Ser-Glu-Lys-Ser-Gln-Thr-Pro-Leu-Val-Thr-Leu (14)

Tyr-Gly-Gly-Phe-Met-Thr-Ser-Glu-Lys-Ser-Gln-Thr-Pro-Leu-Val-Thr-Leu-Phe-Lys-Asn-Ala-Ile-Ile-Lys-Asn-Ala-Tyr (15)

Tyr-Gly-Gly-Phe-Met
(Met-enkephalin) (16)

Tyr-Gly-Gly-Phe-Leu
(Leu-enkephalin) (17)

Fig. 2. Primary structures of endorphins and enkephalins

Shortly afterward partial fragments of β-lipotropin (LPH), e.g. α-, β-, γ-, and δ-endorphin (*12*, *13*, *14*, and *15*) were isolated from pituitary material.

The [Phe^{27}-Gly^{31}]-β_h-endorphin (human) analog of *13*, which was synthesized by C. H. Li et al. [32] on the solid support in 1978, exerts a greater analgesic effect than the natural peptide.

All endorphins have a common initial sequence, which corresponds to the structure of Met-enkephalin.

The morphinomimetic peptides react with the same receptors as the opiate alkaloids and presumably represent the endogenous agonists of these receptors. β-Endorphin, which represents the functionally active molecule, plays a role in the response of the organism to stress stimuli. The analgesic effect in the body can be traced back to the secretion of β-endorphin. Accordingly, acupuncture, for example, activates the central nervous endorphin system and causes an increase in the endorphin concentration, leading to the elimination of sensitivity to pain. Presumably there are endorphinergic systems in the central nervous system (CNS) in which the endorphins assume a neuromodulatory function.

Enkephalins are found in varying amounts in nearly all regions of the nervous system, in the posterior lobe of the pituitary, and in the adrenal cortex. They play a role in pain transmission in that they act as transmitters for the pain-inhibiting neurons in the spinal cord.

Because of their peptide nature, the enkephalins and the endorphins are difficult to put to therapeutic use. The hope that these "brain morphines" would allow analgesia to be separated from the development of addiction and dependence has not yet been realized.

Over 1000 enkephalin derivatives have now been synthesized, and in some of them it has been possible to increase the analgesic effect with respect to enkephalin [33–37]. The enkephalin derivative *18* [38] (Sandoz, FK 33-824), in which enzymatic degradation is blocked, has proved to be strongly analgesically active in animals.

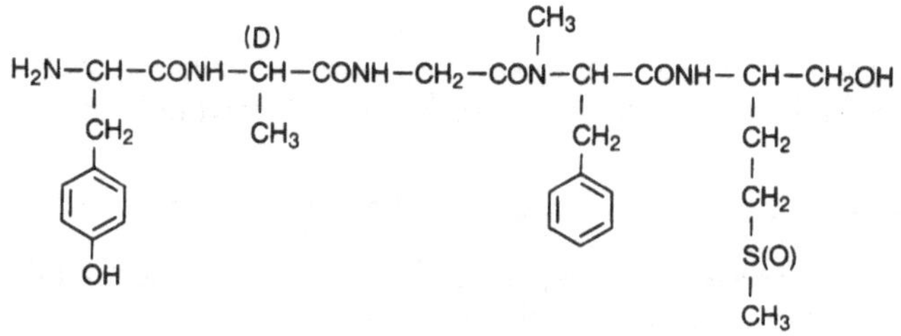

(18)

The analogous enkephalin Tyr-D-Ala-Gly-Paa-Leu [39] only exerts a slight opiate-like effect in comparison with Leu-enkephalin *17*. The exchange of Phe^4 for Paa (β-pyrazinylalanine) leads to a severe loss of activity.

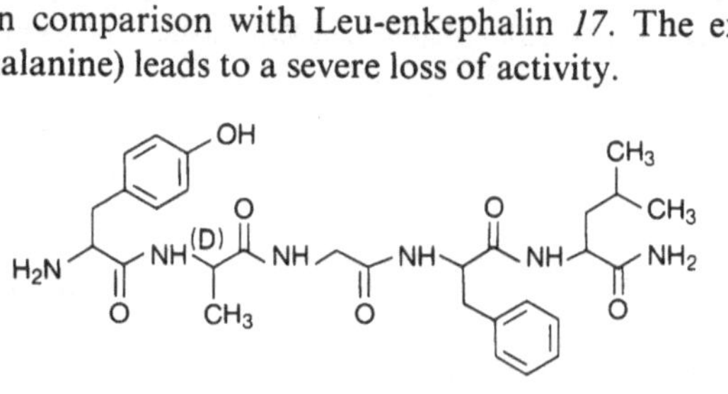

(19)

Further information about the structure-activity relationships was obtained by Schiller et al.[40] with retroinverso modifications of linear enkephalins, e.g. Tyr-D-Ala-gGly-mPhe-Leu-NH_2 *19* (14% Leu-enkephalin activity).

The enkephalin derivatives *20* and *21* from E. Lilly & Co.[41,42] possess analgesic properties..

Tyr-D-Ala-Gly-p-FPhe-D-Phg-NH_2 (20)
(p-FPhe = p-fluorophenylalanine, Phg = α-phenylglycine)

Tyr-D-Ala-Gly-(N^α-allyl)Phe-NH_2 (21)

The cyclohexyl-substituted enkephalins *22* of G. D. Searle & Co.[43] show strong analgesic effects.

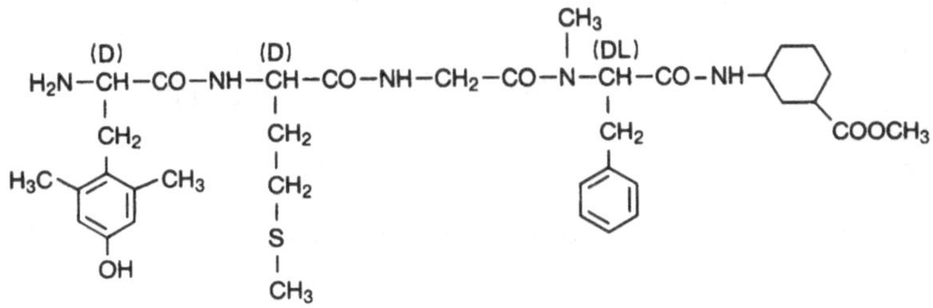

(22)

The dimeric enkephalin *23* synthesized by Imperial Chemical Industries (ICI) Ltd. acts as a selective opiate-receptor antagonist[44].

[Diallyl-Tyr-Gly-Gly-Phe-Leu-NH-$CH_2]_2$ (23)

The N-dihydroxyphosphinylphenylpropionylleucine derivative *24* from the Wellcome Foundation Ltd.[45a] acts as a morphine agonist and an inhibitor of enkephalinase (dipeptidylcarboxypeptidase), which causes hydrolytic cleavage of the Gly^3-Phe^4 linkage of the enkephalin, and hence inactivation.

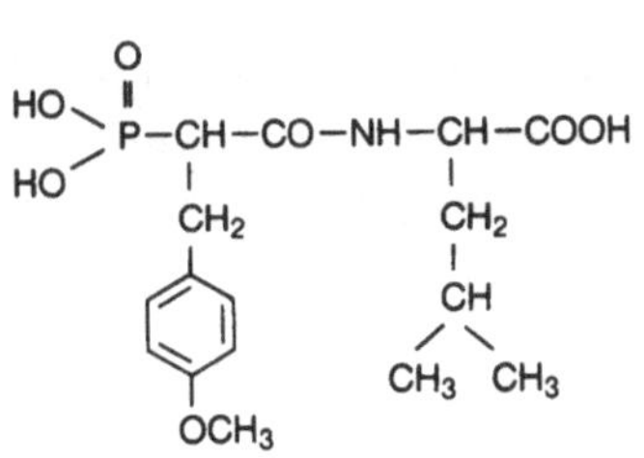

(24)

Schwartz et al.[45b] have recently discussed numerous pharmacological aspects of enkephalin inhibitors such as the analgesic effect, drug design, model predictions about the active center, and the protection of endogenous neuropeptides by peptidase inhibitors.

The racemic inhibitor thiorphan [46] *25* inhibits enkephalinase and selectively supports the analgesic effect of enkephalins.

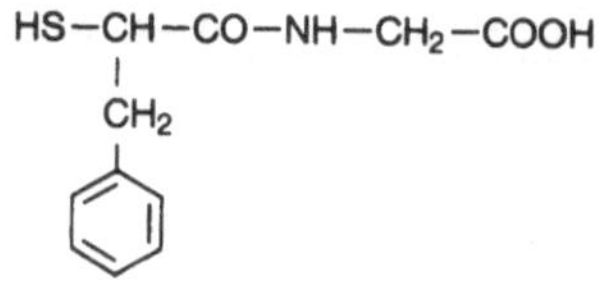

(25)

Similarly, Spatola et al. [47] have synthesized thiomethylene-enkephalin pseudopeptides (Fig. 3), which are stable to proteolytic degradation and exert a biological effect comparable to that of leucine-enkephalin.

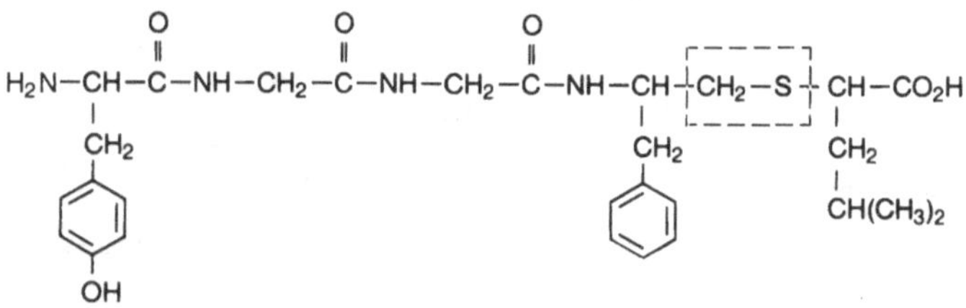

Fig. 3. PheΨ[CH_2S]^{4}Leu-enkephalin; the symbol Ψ[CH_2S] stands in place of the amide linkage (—CO—NH—)

[D-Ala^2]-Met-enkephalinamide [48] Tyr-D-Ala-Gly-Phe-Met-NH_2 and the morphiceptin [49] Tyr-Pro-Phe-Pro-NH_2 have a high morphinomimetic activity (agonist for morphine-μ-receptors).

In 1975 Goldstein et al. [50,51] isolated dynorphin *26*, which is 700 times as effective as Leu-enkephalin, from porcine pituitaries.

Tyr-Gly-Gly-Phe-Leu-Arg-Arg-Ile-Arg-Pro-Lys-Leu-Lys (26)

Dermorphin *27*, isolated from the skin of the frog *Phyllomedusa sauvagei*, exerts a strong analgesic effect and is 700 times as effective as morphine [52,53].

Tyr-D-Ala-Phe-Gly-Tyr-Pro-Ser-NH_2 (27)

Synthetic dermorphin tetrapeptides (small dermorphins), e.g. Tyr-D-Ala-Phe-Gly, and PMRI isomers (partially modified retro-inverso isomers) synthesized by Tomatis et al. [54] are more effective than morphine or dermorphin in the GPI test (guinea-pig ileum test).

2.5 Kyotorphin and Neo-Kyotorphin

The analgesically active dipeptide kyotorphin Tyr-Arg [55], which supposedly causes secretion of Met-enkephalin, was isolated from bovine hypothalamus in 1979 by Takagi et al., as was the pentapeptide neo-kyotorphin (NK) Thr-Ser-Lys-Tyr-Arg [56] in 1982.

The [D-Ser^2]- and [Pro^2]-neo-kyotorphin analogs synthesized by Kitagawa et al. [57] are 10 times as active as native neo-kyotorphin.

The prospects of making pharmacologically more active compounds as well as substances that do not cause dependence by modifying enkephalins, spur the peptide chemist on to greater efforts. As always, the fact that analgesically active peptides are rapidly degraded by enzymes after intracerebroventricular administration in animals represents the main barrier to their therapeutic application.

2.6 Delta-Sleep-Inducing Peptides (DSIP)

DSIP *28*, isolated in 1975 by Monnier et al. [58] from the blood of sleeping rabbits, and the DSIP analogs *29* synthesized by Ivanov et al. [59], produce sleep-like states (δ-slow-wave sleep) after intraventricular infusion (rabbit brain).

Trp-Ala-Gly-Gly-Asp-Ala-Ser-Gly-Glu (28)

Gly-Trp-Ala-Gly-Gly-Asp-Ala-Ser-Gly-Glu (29)

2.7 Releasing and Release-Inhibiting Hormones of the Hypothalamus

The releasing hormones (liberins [60]) and the release-inhibiting hormones (statins) which stimulate the anterior pituitary into hormone production or inhibit release, are low-molecular peptides in comparison with the anterior pituitary hormones and are present in certain areas of the hypothalamus. The hypothalamus exerts an influence on many vital physiological processes in the organism.

2.7.1 Thyrotropin-Releasing Hormone (TRH)

The first releasing hormone to be isolated was TRH pyroGlu-His-Pro-NH_2, in 1969 by Schally et al. [61] and by Guillemin et al. [62] from sheep and porcine hypothalami. The biological activity of the pyroGlu-3-Me-His-Pro-NH_2 synthesized by Burgus et al. [63] exceeds that of natural and synthetic TRH [64] by a factor of 10. TRH regulates the synthesis and the secretion of thyrotropin and prolactin and is used in the diagnosis and therapy of thyroid disorders. The butyrolactone derivative *30a* [65] and pyroGlu-His-3,3-dimethylprolinamide *30b* [66] exhibit CNS activity.

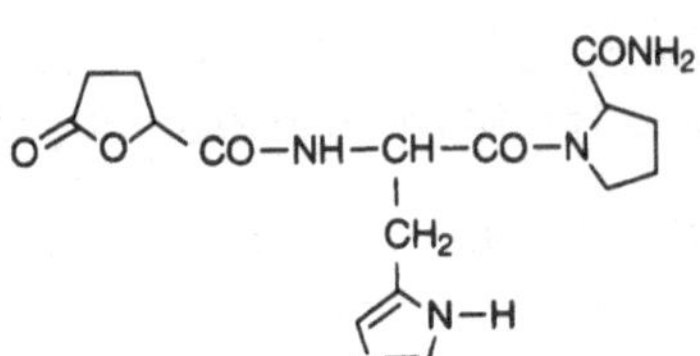

(30a)

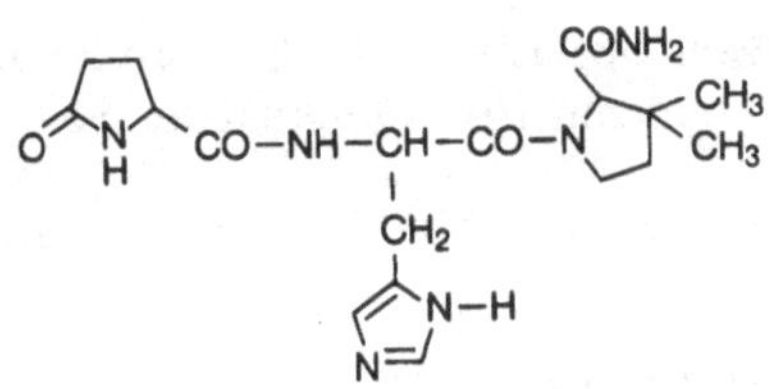

(30b)

2.7.2 Luteinizing Hormone-Releasing Hormone (LH-RH) or Gonadoliberin (Gonadotropin-Releasing Hormone)

LH-RH *31*, isolated in 1971 by Schally et al. [67] and in 1974 by Guillemin from porcine and sheep hypothalamus tissue, possesses LH-releasing and FSH-releasing activity and is available commercially as Lutal® [68].

pyroGlu-His-Trp-Ser-Tyr-Gly-Leu-Arg-Pro-Gly-NH_2 (31)

Stimulation of the secretion of LH (luteinizing hormone) in the female organism triggers ovulation and the formation of the *corpus luteum* responsible for the maintenance of pregnancy. The secretion of FSH (follicle-stimulating hormone, glycoprotein) stimulates the growth and the initial ripening of the follicle in the ovary and therefore sets in motion estrogen biosynthesis.

A broad application of LH-RH analogs is emerging in the fields of contraception and fertility therapy.

Through the synthesis of more than 1000 LH-RH analogs structures have been found that have a biological activity up to 30 times that of the native substance. The compounds of interest at the moment are buserelin *32* [69,70], leuprorelin *33* [71] and nafarelin *34* [72] which, in comparison with native LH-RH, exert a 200-times stronger agonistic effect.

pyroGlu-His-Trp-Ser-Tyr-D-Ser-Leu-Arg-Pro-NH-Et (32)
(D-Ser substituted with Bu^t)

pyroglu-His-Trp-Ser-Tyr-D-Leu-Leu-Arg-Pro-NH-Et (33)

pyroGlu-His-Trp-Ser-Tyr-D-Nal(2)-Leu-Arg-Pro-Gly-NH_2 (34)
Nal(2) = 3-(2-naphthyl)-D-alanine

The growth of testosterone-dependent tumors can be blocked by long-term administration of the three LH-RH analogs. The first therapeutic application of *32*, *33*, and *34* was in the treatment of prostate carcinomas. Buserelin came on the market as a nasal spray (Suprefact®, *32*) and leuprorelin (Carcinil®, *33*) as an injectable product in 1984.

These products effect a blockade of the pituitary gonadotropin secretion and a decrease in the LH-receptors in the Leydig cells of the testes, causing a reduction in the testosterone level (drug castration).

2.7.3 Somatostatin (Growth Hormone-Release-Inhibiting Hormone)

Somatostatin (SST) *35*, a cyclic tetradecapeptide disulfide, was isolated in 1973 by Guillemin from hypothalami. SST has a broad profile of endocrine and gastrointestinal

Ala-Gly-Cys-Lys-Asn-Phe-Phe-Trp-Lys-Thr-Phe-Thr-Ser-Cys (35)
(Cys³–Cys¹⁴ bridged)

effects, i.e. it inhibits not only the secretion of the growth hormone but also the secretion of insulin and glucagon, and therefore plays an important part in the glucose metabolism. In the stomach SST inhibits the secretion of gastrin, hydrochloric acid, and pepsin. In spite of its lack of selectivity and its short half-life, which after parenteral administration is of the order of a few minutes, somatostatin has attracted interest as regards therapeutic uses (treatment of diabetes mellitus, gastric ulcers, and pancreatitis). Synthesis of its analogs has led to compounds that selectively inhibit the secretion of glucagon and insulin, exerting only a slight effect on the release of insulin and an intensified effect on the release of glucagon.

The synthetic modification of SST aimed at achieving a dissociation of effects and the preparation of orally active derivatives, is of practical significance [73-77]. Thus, [D-Trp^8,D-Cys^{14}]-SST *36* preferentially inhibits the liberation of glucagon and the growth hormone (GH); des[Ala^1,Gly^2,Asn^5]-SST and des-Asn^5-SST, on the other hand, inhibit the secretion of insulin, while the secretion of glucagon and GH remain unaffected [78].

Ala-Gly-Cys-Lys-Asn-Phe-Phe-D-Trp-Lys-Thr-Phe-Thr-Ser-D-Cys

(36)

Bicyclic SST analogs *37* from Merck & Co./USA [79a] given i.v. or p.o. cause inhibition of the secretion of insulin, glucagon, and GH.

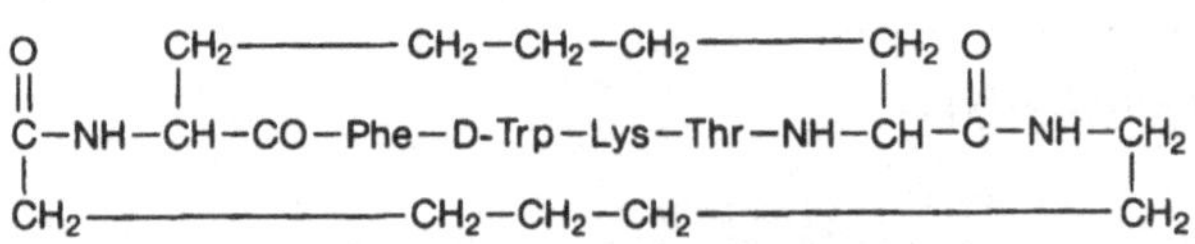

(37)

Retro-enantiomeric [79b] cyclic hexapeptide analogs of SST *38* with a high metabolic stability inhibit the liberation of insulin, glucagon, and the growth hormone [79c].

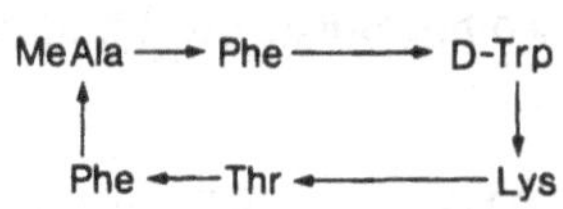

(38)

(MeAla = N-methylalanine)

The cyclic octapeptide SMS 201-995 from Sandoz Ltd./Switzerland *39* [80a] has a longer duration of action than native SST and inhibits the secretion of GH more selectively. Moreover, it enhances the hypoglycemic effect of insulin while simultaneously decreasing glucagon.

D-Phe^1-Cys^2-Phe^3-D-Trp^5

Thr^8(ol)-Cys^7-Thr^6-Lys^5

(39)

Clinical studies with SMS 201-995 [80b] have shown that the growth-hormone concentration in the plasma can be reduced and, in this way, acromegaly (excessive growth of acral regions such as nose, ears, chin, hands, and feet) can be treated by subcutaneous administration.

2.7.4 Corticotropin-Releasing Hormone (CRH or CRF); Melanotropin-Releasing Hormone (MRH); Prolactin-Releasing Hormone (PRH)

The releasing hormone CRH *40*, isolated as a linear peptide-amide with 41 amino acids from ovine hypothalami by Vale et al. [81] in 1981, has certain structural similarities with angiotensinogen. It stimulates the secretion of corticotropin and β-endorphin. In 1985 Morell et al. [82] synthesized ovine CRH by the solid-phase method.

1 5 10 15
Ser-Gln-Glu-Pro-Pro-Ile-Ser-Leu-Asp-Leu-Thr-Phe-His-Leu-Leu-
20 25 30
Arg-Glu-Val-Leu-Glu-Met-Thr-Lys-Ala-Asp-Gln-Leu-Ala-Gln-Gln-
35 40
Ala-His-Ser-Asn-Arg-Lys-Leu-Leu-Asp-Ile-Ala-NH_2 (40)

The N-terminal cyclic hexapeptide of oxytocin and the C-terminal tripeptide of oxytocin were isolated from hypothalami respectively as MRH *41* and melanotropin release-inhibiting hormone *42* (MIH) [5].

Cys-Tyr-Ile-Gln-Asn-Cys (41)

Pro-Leu-Gly-NH_2 (42)

The second-order prohormone oxytocin can be regarded as a precursor of MIH.

Prolactin-releasing hormone (PRH) and prolactin-release-inhibiting hormone (PIH) regulate the formation and the secretion of prolactin in the anterior pituitary [83]. Prolactin itself stimulates the milk secretion from the mammary glands and the growth of these glands. The chemical structures of the two hormones PRH and PIH have not yet been clarified.

2.7.5 Growth-Hormone-Releasing Hormone (GH-RH or GRF) or Somatocrinin

In 1982 Guillemin et al. [84] and Rivier et al. [85] isolated 3 peptides *43a* with GRF activity (stimulation of the secretion of growth hormone) from human pancreatic tumor cells, and synthesized them by the solid-phase method:

GRF(1–44)-NH_2, GRF(1–40)-OH and GRF(1–37)-OH (43a)

H-Tyr-Ala-Asp-Ala-Ile-Phe-Thr-Asn-Ser-Tyr-Arg-Lys-Val-Leu-Gly-
Gln-Leu-Ser-Ala-Arg-Lys-Leu-Leu-Gln-Asp-Ile-Met-Ser-Arg-Gln-
Gln-Gly-Glu-Ser-Asn-Gln-Glu-Arg-Gly-Ala-Arg-Ala-Arg-Leu-NH_2
37 40 44
(43b)

Other syntheses by H. Yajima et al. [86] have become known. *43b* has been used therapeutically, for example in wound healing.

2.8 Proteohormones of the Pituitary (Hypophysis)

As a rule, a hypothalamic hormone should always control a hormone in the anterior pituitary. The pituitary hormones (e.g. follicle-stimulating hormone, prolactin, or thyrotropin) are then transported through the bloodstream to the secondary target organs, where, for example, they stimulate the production of corticosteroids in the adrenals or the formation of thyroxine in the thyroid.

2.8.1 Somatotropin (Growth Hormone)

The growth hormone (STH) or the human growth hormone [87] (HGH), a linear peptide hormone made up of 191 amino acids with 2 intrachain disulfide bridges, is formed under the control of somatostatin and influences the maturation process during the growth period (e.g. the increase in protein substance and in height).

A substance isolated from human pituitary (e.g. Asellacrin®) is available for the treatment of pituitary dwarfism in which there is a confirmed STH deficiency.

On the other hand, excessive secretion of HGH in the growing years leads to gigantism. HGH is also used in cases of muscular dystrophy, bone decalcification (osteoporosis), and hemorrhagic gastric ulcers.

The synthesis of the human growth hormone by DNA recombination is of major significance in view of the difficulties of total synthesis.

2.8.2 Corticotropin (ACTH)

Adrenocorticotropic hormone (Fig. 4) stimulates the cells of the adrenal cortex into the secretion and production of steroid hormones. Conversely, the pituitary secretion of ACTH is inhibited by the adrenal hormones via a feedback mechanism.

```
        Ser-Tyr-Ser-Met-Glu-His-Phe-Arg-Trp-Gly-Lys-Pro-Val-
        1           5                   10
human   Gly-Lys-Lys-Arg-Arg-Pro-Val-Lys-Val-Tyr-Pro-Asn-Gly-
            15                  20                  25
        Ala-Glu-Asp-Glu-Ser-Ala-Glu-Ala-Phe-Pro-Leu-Glu-Phe
                    30                  35
porcine                 -Leu-Ala-Glu-
bovine                  -Ser-Ala-Gln-
```

Fig. 4. Amino acid sequences in human, porcine, and bovine ACTH

Since 1956 more than 150 partial sequences and analogs [88, 89] have been synthesized, mainly with chain lengths of 1–16 or 1–28. The first total synthesis of porcine ACTH was described in 1963 by Schwyzer et al. [90].

Within the framework of ACTH synthesis a broad knowledge of the structure-activity relationships has been acquired concerning the influence of the chain length on biological activity.

The ACTH sequence can be formally subdivided into various sections of differing biological significance. The N-terminal section 1–10 represents the active center, while sequence 11–18 is responsible for receptor binding. The C-terminal section 25–39 contains the hormonal information for species specificity and for antigenicity.

The N-terminal tetracosapeptide of ACTH (Synacthen) finds therapeutic application in the treatment of arthrorheumatism, bronchial asthma, and nephroses.

2.8.3 Oxytocin and Vasopressin

The actual site of formation of oxytocin and vasopressin is in the hypothalamus, from which the two peptides are carried to the posterior pituitary bound to neurophysins (transport proteins) and stored. The structure and synthesis of oxytocin *44* and vasopressin *45* were worked out by du Vigneaud et al. [91].

Cys-Tyr-Ile-Gln-Asn-Cys-Pro-Leu-Gly-NH_2 (44)

Cys-Tyr-Phe-Gln-Asn-Cys-Pro-Arg-Gly-NH_2 (45)

These peptide hormones are among the best-researched active peptide substances. More than 350 oxytocin and vasopressin analogs have been reported in the literature.

Oxytocin causes contraction of the uterine smooth muscles and stimulates milk ejection in the lactating glands.

The vasopressins cause reabsorption of water by increasing renal permeability, thus concentrating the primary urine. If the vasopressin level is too low, the reabsorption of water is no longer ensured, so that large quantities of urine of low specific gravity are excreted (water diuresis = diabetes insipidus). With high doses of vasopressin the blood pressure and the intestinal peristalsis are increased.

Oxytocin is used in obstetrics to induce labor, e.g. to maintain the uterine contractions during birth, and to promote the evacuation of milk. The most important therapeutic use of vasopressin is based on its antidiuretic effect in diabetes insipidus (e.g. 1-desamino-D-Arg^8-vasopressin as a nasal spray [92]).

2.9 Carnosine

The dipeptide carnosine β-Ala-His is found in skeletal muscle in relatively large amounts, and functions presumably as a neurotransmitter. Clinical studies have shown that carnosine accelerates wound healing [93].

2.10 Invertebrate Neuropeptide Hormones

The invertebrate peptides PCH (pigment-concentrating hormone *46*) and AKH I (adipokinetic hormone *47*) were the first neuropeptides from invertebrates to have their structure and synthesis described by L. Josefsson in 1983 [94].

pyroGlu-Leu-Asn-Phe-Ser-Pro-Gly-Trp-NH_2 (46)

pyroGlu-Leu-Asn-Phe-Thr-Pro-Asn-Trp-Gly-Thr-NH_2 (47)

Both peptides play a role in the mechanism of color adaptation in insects.

2.11 DBI Peptides (Diazepam-Binding Inhibitors)

Guidotti and Ferrero [95] isolated from human and rat brain extracts neuroactive peptides that interact with the receptors at which benzodiazepines (e.g. Valium and Librium) induce biological effects. The DBI peptide *48* ("anxiety peptide"), an endogenous ligand of the benzodiazepine receptor, causes anxiety, in contrast to the benzodiazepines.

Gln-Ala-Thr-Val-Gly-Asn-Val-Asn-Thr-Asp-Arg-Pro-Gly-Leu-
Leu-Asp-Leu-Lys (48)

Neuropeptide research will in all probability allow the development of new active substances that are more effective, more specific, and safer than the psychopharmaceuticals in current use.

3 Gastrointestinal Peptides (Peptides of the Stomach, Intestine, and Pancreas)

Peptide hormones (aglandular hormones), whose action makes possible the secretory processes necessary for the normal course of the digestive process, are formed in the gastric and intestinal mucosae and in the excretory pancreatic tissue (islets of Langerhans).

3.1 Secretin, Glucagon, VIP, PHI, and GIP (Fig. 5)

Secretin, isolated from the duodenal mucosa by Jorpes and Mutt [96] in 1961, and 4 years later found to be a linear heptacosapeptide, stimulates the pancreas to produce a bicarbonate-containing secretion. Bodanszky et al. [97], Ondetti et al. [98], and Wünsch et al. [99] have synthesized secretin derivatives. Another synthesis, by Uchiyama et al. [100] using fragment condensation on a large scale, led within a short time to highly-purified secretin.

Glucagon, which was isolated from porcine pancreas in 1953 by Staub et al. [101] and structurally clarified by Bromer et al. [102] in 1956, exerts hyperglycemic (insulin-antagonistic) and positive inotropic effects. Glucagon is used therapeutically in hypoglycemic states resulting from insulin overdosage, heart failure, or in cases of β-blocker overdoses. Wünsch et al. [103] and R. B. Merrifield [104] have published extensive works on glucagon.

As linear polypeptides, VIP (vasoactive intestinal polypeptide) [105], PHI (peptide HI: H = N-terminal His, I = C-terminal Ile) [106,107], and GIP (gastric inhibitory polypeptide consisting of 43 amino acids) [108] are structurally similar to secretin and glucagon. VIP and PHI act as vasodilators, exert hyperglycemic effects, and affect the smooth muscle of the gallbladder. GIP completely blocks gastric secretion.

	1 2 3 4 5 6 7 8 9 10 11 12 14 14 15
VIP	His-Ser-Asp-Ala-Val-Phe-Thr-Asp-Asn-Tyr-Thr-Arg-Leu-Arg-Lys-
Secretin	His-Ser-Asp-Gly-Thr-Phe-Thr-Ser-Glu-Leu-Ser-Arg-Leu-Arg-Asp-
Glucagon	His-Ser-Gln-Gly-Thr-Phe-Thr-Ser-Asp-Tyr-Thr-Lys-Tyr-Leu-Asp-
GIP	Tyr-Ala-Glu-Gly-Thr-Phe-Ile-Ser-Asp-Tyr-Ser-Ile-Ala-Met-Asp-
PHI	His-Ala-Asp-Gly-Val-Phe-Thr-Ser-Asp-Phe-Ser-Arg-Leu-Leu-Gly-

	16 17 18 19 20 21 22 23 24 25 26 27 28 29
VIP	Gln-Met-Ala-Val-Lys-Lys-Tyr-Leu-Asn-Ser-Ile-Leu-Asn-NH_2
Secretin	Ser-Ala-Arg-Leu-Gln-Arg-Leu-Leu-Gln-Gly-Leu-Val-NH_2
Glucagon	Ser-Arg-Arg-Ala-Gln-Asp-Phe-Val-Gln-Trp-Leu-Met-Asp-Thr
GIP	Lys-Ile-Arg-Gln-Gln-Asp-Phe-Val-Asn-Trp-Leu-Ala-Gln-Gln ... Gln
PHI	Gln-Leu-Ser-Ala-Lys-Lys-Tyr-Leu-Glu-Ser-Leu-Ile-NH_2

Fig. 5. Comparison of the amino acid sequences of secretin, glucagon, VIP, PHI, and GIP

3.2 Gastrin, Cholecystokinin-Pancreozymin (CCK-PZ) and Galanin

Gastrins I and II (Tyr^{12}-O-sulfated gastrin I) *49*, whose structure was elucidated in 1964 by Gregory et al. [109], are formed in the mucosal lining of the gastric antrum and cause strong stimulation of the secretion of hydrochloric acid in the stomach and of enzyme secretion in the pancreas. The C-terminal tetrapeptide of Trp-Met-Asp-Phe-NH_2 already has the physiological properties of the natural hormone [110]. The solid-phase method has proved suitable for the synthesis of gastrin peptides according to Sheppard [111]. Recently Merrifield et al. [112] synthesized human gastrin I on benzhydrylamine resin using the "low-high HF" technique.

pyroGlu-Gly-Pro-Trp-Leu-Glu-Glu-Glu-Glu-Glu-Ala-Tyr-Gly-
Tyr-Gly-Trp-Met-Asp-Phe-NH_2 (49)
(human)

CCK-PZ *50* [113–116], whose C-terminal pentapeptide sequence is identical with that of gastrin, is discharged into the blood from certain intestinal cells and reaches the gallbladder, where it causes the bile to discharge into the intestine, thus promoting digestion. Cholecystokinin also functions as a cerebral neurotransmitter.

Lys-Ala-Pro-Ser-Gly-Arg-Val-Ser-Met-Ile-Lys-Asn-Leu-Gln-Ser-Leu-
Asp-Pro-Ser-His-Arg-Ile-Ser-Asp-Arg-Asp-Tyr-Met-Gly-Trp-Met-Asp-
SO_3H (on Tyr)
Phe-NH_2 (50)

Castro et al. [117] have recently been able to show with the aid of structure-activity studies that tripeptide and tetrapeptide derivatives such as Boc-β-Ala-Trp-Met-Asp-NH_2 *51a* from the C-terminal region of gastrin without phenylalanine as a gastrin antagonist inhibit in vivo the gastrin-stimulated acid secretion.

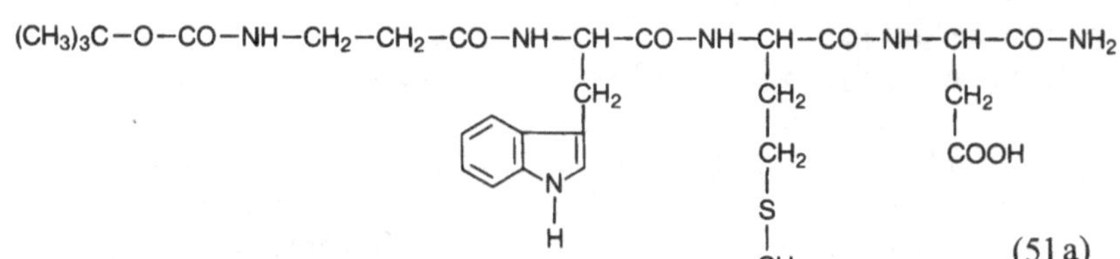

(51a)

On the other hand, pseudopeptide analogs of the C-terminal tetrapeptide of gastrin [117b], such as (*tert.*butyloxycarbonyl)-L-tryptophyl-Ψ(CH_2—NH)-L-leucyl-L-aspartyl-L-phenylalaninamide *51b*, in which the amide linkage is replaced by the isosteric modification CH_2—NH, are potent agonists of acid secretion.

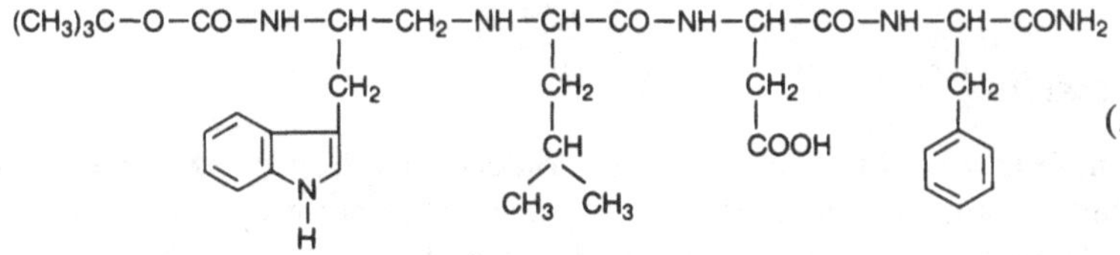

(51b)

Galanin *52*, which was isolated from porcine intestine by Tatemoto et al. [118] in 1983, causes contraction of the smooth muscle and mild hyperglycemia in dogs.

Gly-Trp-Thr-Leu-Asn-Ser-Ala-Gly-Tyr-Leu-Leu-Gly-Pro-His-
Ala-Ile-Asp-Asn-His-Arg-Ser-Phe-His-Asp-Lys-Tyr-Gly-Leu-Ala-NH_2 (52)

3.3 Gastrin-Releasing Peptide (GRP), Bombesin, and Motilin

GRP [119] *53* was isolated from porcine mucosae and causes gastrin secretion. The C-terminal decapeptide of GRP is, with the exception of His^{20}, identical with that of bombesin *54*, and also coincides with the sequence of the decapeptide neuromedin C [119b,c] [Gly^{18}-Met^{27}]-GRP. Neuromedin C, a porcine spinal-cord peptide that can also be regarded as a bombesin-like peptide, exerts stimulating effects on rat uterine smooth muscle and functions as a neuromediator in the neural communication systems of mammals.

1 10
Ala-Pro-Val-Ser-Val-Gly-Gly-Gly-Thr-Val-Leu-Ala-Lys--Met-
18 27
Tyr-Pro-Arg-Gly-Asn-His-Trp-Ala-Val-Gly-His-Leu-Met-NH_2 (53)

1 5
pyroGlu-Gln-Arg-Leu-Gly-Asn-Gln-Trp-Ala-Val-Gly-His-Leu-
14
Met-NH_2 (54)

Bombesin *54* [120, 121], isolated from frog skin, is thought to have antidiuretic and antihypertensive properties. It stimulates the secretion of gastrin, pancreatic and gastric secretion, and causes contraction of the gallbladder.

Recently Moody et al. [122] discovered that the C-terminal partial sequence of bombesin and bombesin-like peptides (BLPs) can function as autocrine growth factors in human small-cell lung cancer (SCLC) cell lines.

Motilin *55* (gastric motor activity-stimulating polypeptide), isolated from porcine intestine by Mutt and Brown [123], stimulates gastric motility and pepsin secretion.

Phe-Val-Pro-Ile-Phe-Thr-Tyr-Gly-Glu-Leu-Gln-
Arg-Met-Gln-Glu-Lys-Glu-Arg-Asn-Lys-Gly-Gln (55)

Wünsch et al. [124] synthesized the fragments 9–22 from [Nle^{13}, Glu^{14}]-motilin and [Leu^{13},Glu^{14}]-motilin.

3.4 Insulin [5, 21, 89, 125]

Insulin (Fig. 6), which belongs to the older-generation polypeptides was discovered by Banting and Best in 1921 and its primary structure was elucidated by Sanger in 1955. In view of the very diverse syntheses and the numerous biological publications on this substance, only a general summary is possible here.

The first total synthesis of A- and B-chains and their combination into insulin was performed by Zahn et al. [126] in 1963.

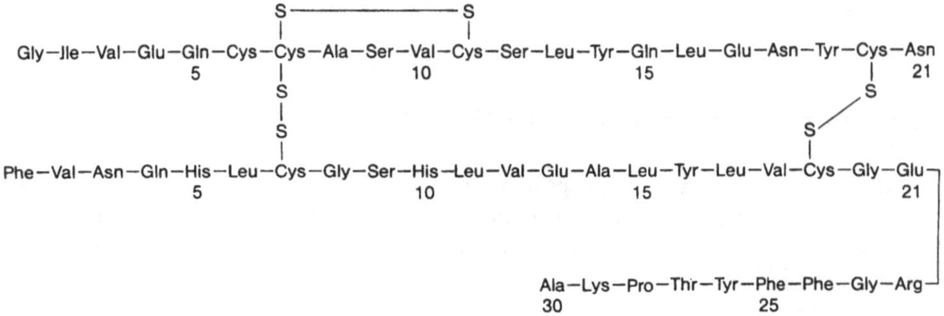

Fig. 6. Structure of bovine insulin

Insulin does not display any pronounced organ specificity, in fact numerous metabolic processes in the liver, muscle, and fat cells are insulin-dependent.

Under the influence of insulin the permeability of the cell membranes of many organs and tissues is increased, and the material transport from the extracellular space into the liver, fat, and muscle cells in promoted. The accelerated influx of glucose into these cells due to insulin leads to an increased glucose degradation. The glucose level is therefore reduced by insulin. Because of the great influence of insulin on the glucose transport through the plasma membranes of fat and muscle cells, it is presumed that the insulin receptors [127] on the surface of these cells are the hormone's main site of action.

Insulin affects the protein metabolism by an increased uptake of amino acids as a result of the increased permeability of the cell membranes.

Diabetes mellitus is the result of too little insulin being released from the pancreatic β-cells or of a decrease in the number of active insulin receptors in the target tissue.

Many preparations are available for the treatment of diabetics with insulin. Rittel et al. [128] described total synthesis of human insulin in 1974. Since that time new ways of obtaining insulin have been developed, namely the E. Lilly & Co. biotechnological synthesis by genetically modified microorganisms [129] (DNA recombination) and the trypsin-catalyzed conversion of porcine into human insulin by exchanging the C-terminal Ala^{30} of the B-chain for threonine (Novo Research Institute/Denmark [130]).

4 Peptides of Immunological Importance

Peptide chemists are becoming increasingly interested in the synthesis of immunosuppressive and immunostimulating peptides, which are discussed below.

4.1 Peptides of the Thymus

The thymus, a gland situated under the sternum (primary lymph organ), occupies a central position in the immune system. Among its functions are the stimulation of the immune responses and cell differentiation. If foreign substances invade the organism, the body reacts defensively via its immune system.

The carriers of the immune system are the leucocytes, the immune cells [21]. These are either phagocytotic cells (e.g. granulocytes and macrophages) or lymphocytes, namely T-lymphocytes, which differentiate further in the thymus, and B-lymphocytes, which stem from the parent cells of the spinal cord.

In the course of an immune response the finally differentiated plasma cells, whose function is antibody synthesis, are formed from the B-lymphocytes under the influence of the T-lymphocytes and the thymus peptides.

4.1.1 Thymopoietin Derivatives (TP)

Thymopoietin, a calf thymus polypeptide with 49 amino acids, causes selective T-cell differentiation and is therefore responsible for the cell-bound immune reaction. Consideration of smaller peptides having the same properties with respect to the immune system led to the synthetic thymopentin Arg-Lys-Asp-Val-Tyr [131,132] and derivatives such as Lys-Lys-Tyr-Phe-Arg [133].

4.1.2 Thymosin α_1

Thymosin α_1 *56*, isolated by A. L. Goldstein et al. [134] from bovine thymus glands, is important for the development of thymus-dependent T-lymphocytes.

CH_3CO-Ser-Asp-Ala-Ala-Val-Asp-Thr-Ser-Ser-Glu-
Ile-Thr-Thr-Lys-Asp-Leu-Lys-Glu-Lys-Lys-Glu- (56)
Val-Val-Glu-Glu-Ala-Glu-Asn

Thymosin α_1 has been synthesized by several research groups by the solid-phase method [135,136] as well as by conventional synthetic techniques [137,138].

4.1.3 Serum Thymic Factor (FTS or Thymulin)

Bach et al. [139] isolated thymulin *57*, which is responsible for T-cell differentiation, from porcine serum.

Pyroglu-Ala-Lys-Ser-Gln-Gly-Gly-Ser-Asn (57)

Numerous FTS analogs that exhibit biological effects in the differentiation of T-lymphocytes have now been synthesized [140]. Information is already available about the possibility of using the peptide derivatives thymosin α_1 and FTS therapeutically in the treatment of disorders of the immune system (e.g. in patients with congenital T-cell deficiency or chronic bacterial infections).

4.2 Tuftsin

Tuftsin *58*, isolated from the protein leukokinin in 1970 by Najjar et al. [141] by enzymatic cleavage of a γ-globulin fraction, stimulates phagocytosis.

Thr-Lys-Pro-Arg (58)

Because of the action of tuftsin on macrophages and granulocytes, its therapeutic use is being tested in various infectious diseases [142].

4.3 Cyclosporines [143]

Cyclosporines (Fig. 7) are cyclic undecapeptides composed of 11 amino acids — some of them N-methylated — in which the amino acid (4R)-4-[(E)-2-butenyl]-4,N-dimethyl-L-threonine (MeBmt) *59* is a characteristic component and plays a considerable role in the biological activity.

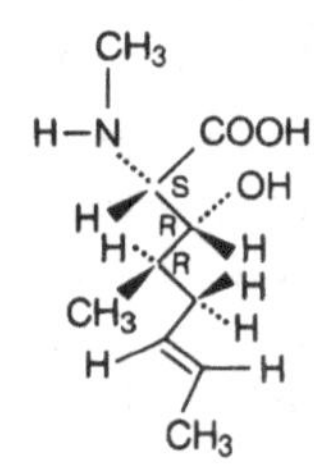

(59)

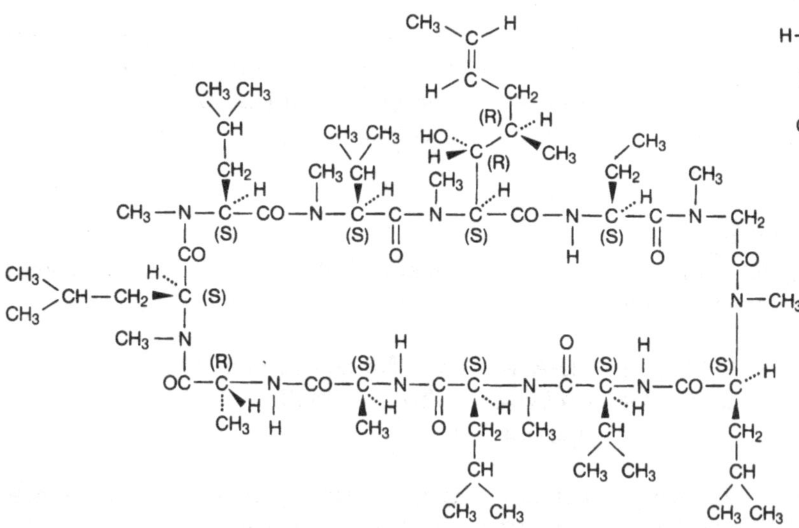

Fig. 7. Structure of cyclosporin A

Cyclosporin A, which has been used as an immunosuppressant with fungicidal and anti-inflammatory properties in bone marrow and organ transplants and in autoimmune diseases since 1983, was isolated from a fungal culture (*Tolypocladium*) [144].

The structure of cyclosporine was established in 1976 by Petcher et al.[145] and its total synthesis was accomplished in 1984 by Wenger[146,147].

4.4 Muramyl Peptides (Glycopeptides)

The muramyl peptides (MP) of the bacterial cell walls are polymeric chain molecules composed of N-acetylglucosamine and N-acetylmuramic acid, each carrying a peptide side chain and crosslinked with one another (Fig. 8). The N-terminal D-Ala residues of the side chains are bound covalently to the peptide side chains of neighboring polysaccharide chains via pentaglycine units.

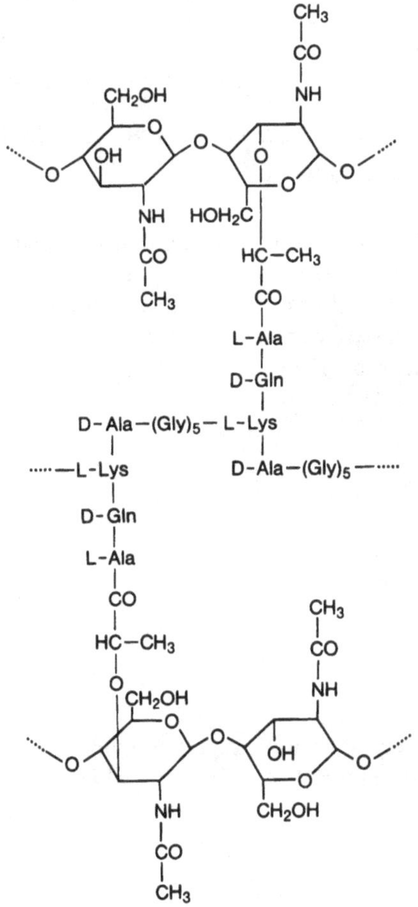

Fig. 8. Basic structural unit of a peptidoglycan of the bacterial cell wall of *Staphylococcus aureus*

Ciba-Geigy AG[148] have synthesized the N-acetylmuramyl-L-alanyl-D-isoglutamine-2-(hexadecyloxyhydroxyphosphoryloxy)-ethylamide *60*, which has immunopotentiating properties.

Muramyl peptides can be used as adjuvants in combination with vaccines (increasing the humoral and cellular immunity) or with antibiotics (increasing the antibacterial effect).

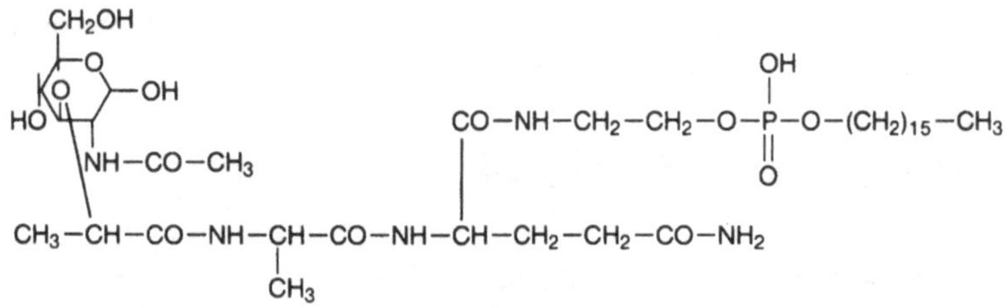

(60)

The mitogenic pentapeptide *61* S-[2,3-bis(palmitoyloxy)-(2 R,S)-propyl]-N-palmitoyl-Cys-Ser-Ser-Asn-Ala (TPP) of Jung et al. [149], which is a component of the lipoprotein from the outer membrane of *Escherichia coli*, also acts as a potent adjuvant and B-lymphocyte activator.

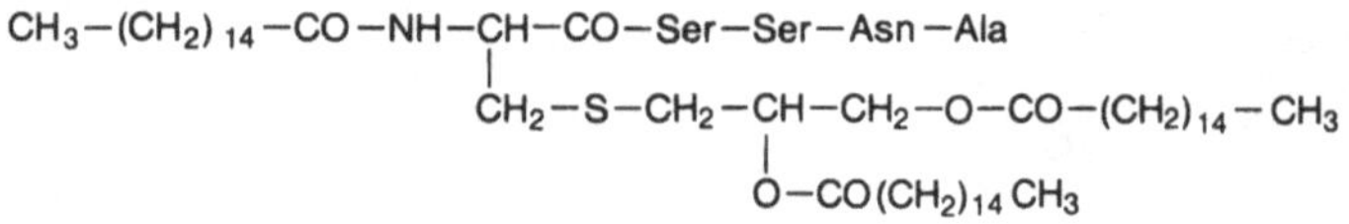

(61)

The biological effect of the muramyl derivatives is based on the fact that the tissue macrophages are activated and form the first line of defense of the immune system against invaders (phagocytosis of the antigen). In addition, the lymphatic system is activated, i.e. the B- and T-lymphocytes are stimulated (specific immune response) [150]. Protected asparagine glycopeptides *62* were made by Kunz et al. [151] by increasing the length of the N-terminal peptide chain.

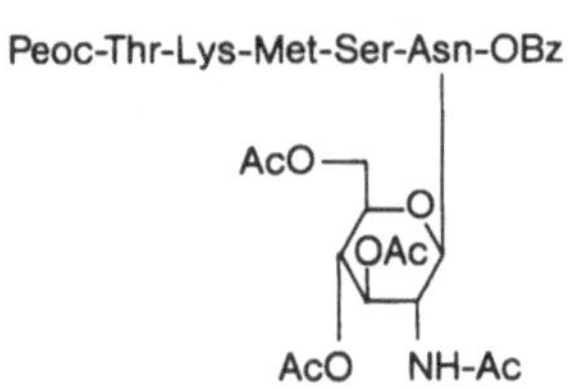

(62)

Ac: Acetyl
Bz: Benzyl
Peoc: 2-(Triphenylphosphonio) ethoxycarbonyl

The N-acetylglucosaminyl-1,6-anhydro-N-acetylmuramylalanylglutamyldiaminopimelylalanine (NAG-1,6-anhydro-NAM-Ala-Glu-dap-Ala, Fig. 9), which was

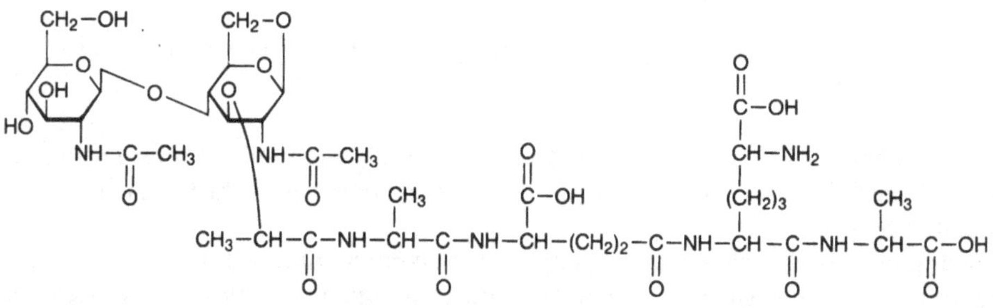

Fig. 9. NAG-1,6-anhydro-NAM-Ala-Glu-dap-Ala

isolated from human urine as sleep-promoting factor S by J. M. Krueger et al.[152], exerts a somnogenic effect, i.e. the slow-wave sleep phase (SWS) is prolonged.

Numerous analogs of the muramyl peptide factor[153, 154] have been synthesized with a view to finding endogenous hypnotics that do not lead to dependence.

4.5 Peptide Vaccines

Many research groups are currently engaged in producing antibodies directed against a synthetic segment of a protein[155, 156].

Peptides as immunogens that trigger the formation of antibodies of effector cells (killer cells) in higher organisms could inaugurate a new generation of vaccines.

Antibodies of this type are formed if the synthetic peptide used as the immunogen is the same as a segment of a protein (antibody) that lies on the surface in its native form, and whose native conformation can in part be adopted by the peptide (antigen).

Active immunization occurs if the antibodies formed protect the organism from viruses or toxins. It is hoped that synthetic peptides triggering the formation of antibodies will find use as potential vaccines.

The future importance of peptide vaccines lies in the fact that one could replace inactivated or attenuated microbial pathogens or toxins, which are high-molecular and therefore difficult to characterize and standardize, by highly specific synthetic peptides. Emini et al.[157] have synthesized oligopeptides that prime the rabbit immune system and are effective against *poliovirus*. The amino acid sequence of the peptide vaccines *63* and *64* originate in the poliovirus VP_1 protein.

Tyr-Gly-Gly-Ser-Thr-Thr-Asn-Lys-Asp-Lys-Gly (63)
Tyr-Gly-Gly-Asp-Asn-Pro-Ala-Ser-Thr-Thr-Asn-Lys-Asp-Lys (64)

Synthetic peptide vaccines against the influenza virus, for example, could become of major therapeutic interest.

4.6 FK-156 and FK-565

D-Lactoyl-L-alanyl-γ-D-glutamyl-(L)-meso-diaminopimelyl-(L)-glycine (FK-156) *65*, isolated from *Streptomyces olivaceogriseus*[158] and subsequently synthesized[159], is an adjuvant-active and immunostimulating peptide[160].

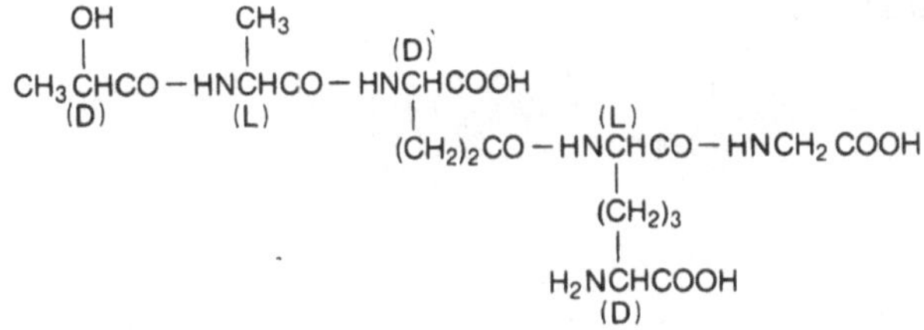

(65)

Heptanoyl-γ-D-glutamyl-(L)-meso-diaminopimelyl-(D)-alanine (FK-565) *66*, a fatty acid derivative of the bacterial cell wall peptidoglycans, and also FK-156 analogs [161, 162], suppress the growth of tumors when administered subcutaneously and orally and at the same time are effective against bacterial infections.

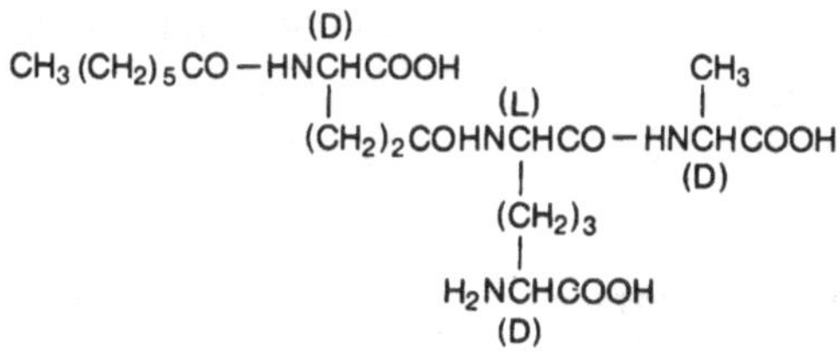

(66)

4.7 Chlamydocins and Peptidylacivicins

Chlamydocin *67*, isolated as a cyclic tetrapeptide from culture broths of *Diheterospora chlamydosporia* by Closse et al. [163] in 1974, has 100 times the cytostatic activity of actinomycin D with respect to inhibition of cell growth in the mouse. Numerous chlamydocin derivatives have been synthesized by D. H. Rich et al. [164].

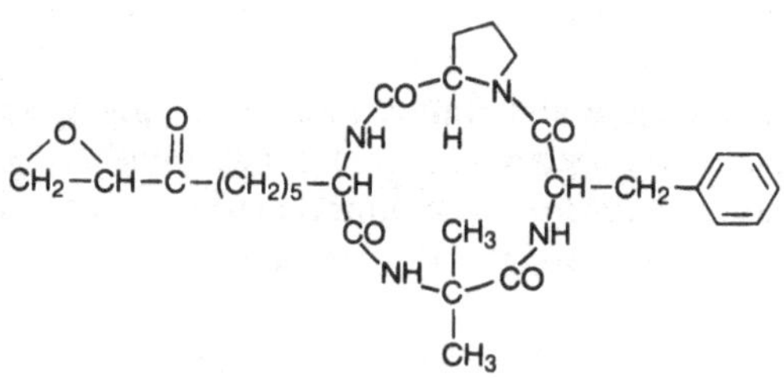

(67)

The characteristic basic unit of the cyclopeptide is S-2-amino-S-9,10-epoxy-8-oxodecanoic acid. U. Schmidt et al. have reported the synthesis of chlamydocin and epichlamydocin [165]. One of the main problems in the treatment of cancer is always the high toxicity of cytostatics and their low selectivity against malignant cells. Attempts to develop prodrugs that are only activated in the vicinity of a tumor by a tumor-enzyme inhibition have been made with peptidylacivicins. Tumor cells contain high levels of plasmin-activator and therefore high protease-plasmin levels. According to Katzenellenbogen et al. [166], with the aid of AT-125 peptide *68* it should be possible to achieve the full effect of plasmin-activated prodrugs only in the cancer cells.

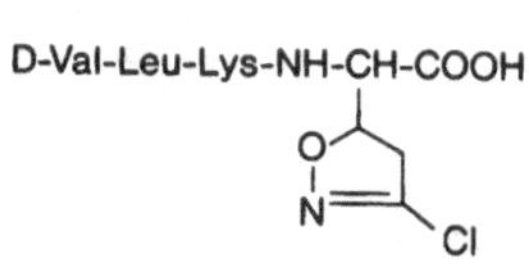

(68)

4.8 Bestatin

Bestatin *69* [(2S,3R)-3-amino-2-hydroxy-4-phenylbutanoyl]-L-leucine, which can either be obtained from culture broths of *Streptomyces olivoreticuli* as per H. Umezawa et al. [167] or by synthesis [168], can be used in the treatment of malignant skin tumors.

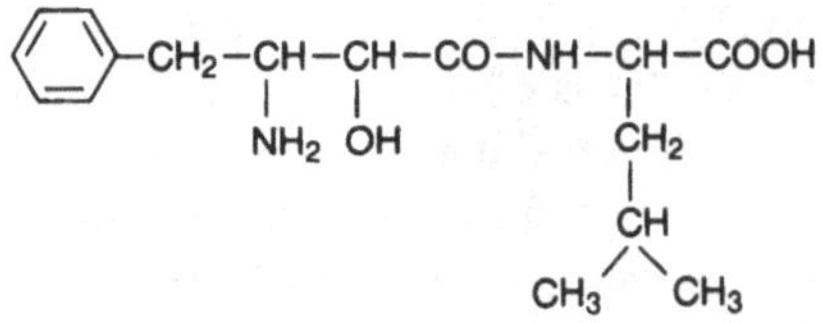

(69)

5 Peptides of the Renin-Angiotensin System

5.1 Proteolytic Cascade of the Renin-Angiotensin System [169, 170]

Since the renin-angiotensin system is involved in regulation of the blood pressure, antihypertensive substances are encountered among the peptides that exert an effect on this system. The mechanism of the renin-angiotensin system is illustrated in Fig. 10.

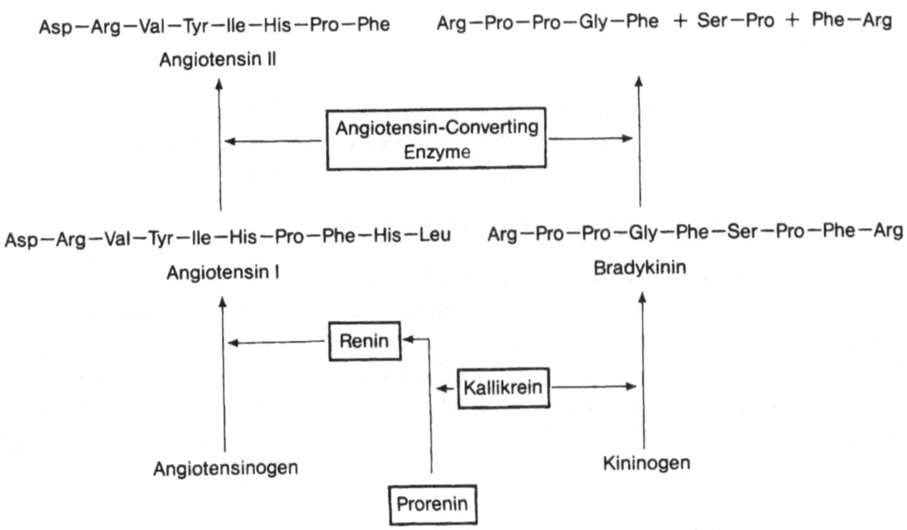

Fig. 10. Processing of angiotensinogen and kininogen

When the blood pressure is reduced, or during sympathetic stimulation, renin is secreted from the juxtaglomerular kidney cells. This enzyme liberates a biologically inactive decapeptide, angiotensin I, from the renin substrate (angiotensinogen) produced in the liver, with cleavage of a Leu-Leu or a Leu-Val bond (human). Under the influence of a peptidyldipeptide hydrolase likewise present in plasma (angiotensin-converting enzyme = ACE), the biologically active angiotensin II is formed from angiotensin I by splitting-off of the C-terminal dipeptide His-Leu (Fig. 10). Angiotensin II has a contracting effect on the vascular smooth muscle and is the most powerful vasoconstrictor known.

5.2 Angiotensin II Antagonists and ACE Inhibitors

Inhibition of the action of angiotensin II *70* has been achieved with many angiotensin I and II analogs (Fig. 11), in which, on the basis of structure-activity relationships, an essential function is attributed to the Phe^8 residue in stimulation of the receptor, while Tyr^4 and His^6 residues are thought to be important in the binding to the receptor. These angiotensin II-blockers (*70–75*) prevent the action of the effector peptide of the renin-angiotensin system on the target organ.

Saralasin *71* is currently in use as an angiotensin-receptor antagonist in the treatment of hypertension, despite its partial agonistic effect. The disadvantages of *71* lie in the fact that it is administered i.v. and in its short biological half-life.

Asp-Arg-Val-Tyr^4-Ile-His^6-Pro-Phe^8	(70)
Sar-Arg-Val-Tyr-Val-His-Pro-Ala	(71)
Sar-Arg-Val-Tyr-Ile-His-Pro-Ile [171,172]	(72)
Sar-Arg-Val-Tyr-Ile-His-Pro-Thr [173]	(73)
Sar-Arg-Val-Tyr-Ile-His-Pro-Ala [174]	(74)
Asp-Arg-Val-Tyr-Ile-His-Ala-Phe [175]	(75)

Fig. 11. Angiotensin II antagonists

The teprotide BPP_{9a} (bradykinin-potentiating peptide) *76* and analogous BPP peptides [170], which have been isolated from snake venom, are taken up by ACE in competition with the substrate angiotensin I with far greater affinity.

pyroGlu-Trp-Pro-Arg-Pro-Gln-Ile-Pro-Pro (76)

Converting enzyme (ACE) inhibitors [176] likewise prevent the formation of angiotensin II and are used in the treatment of renal and essential hypertension. Examples of orally active ACE-inhibitors are: (2)-1-[(2S)-3-[N-(S)-mercapto-2-methylpropanoyl]proline [170] (captopril *77*), 1-[N-(S)-1-carboxy-3-phenylpropyl]-L-alanyl-L-proline-1′-ethyl ester [177] (enalapril *78*), and 2-[N-[(S)-1-ethoxycarbonyl-3-phenylpropyl]-L-alanyl]-(1S,3S,5S)-2-azabicyclo[3.3.0]octane-3-carboxylic acid [178] (Hoe 498; *79*).

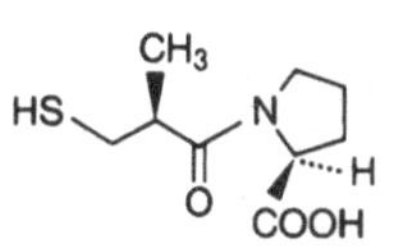

(77)

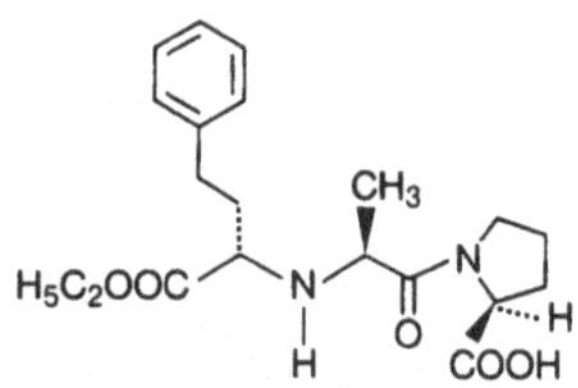

(78)

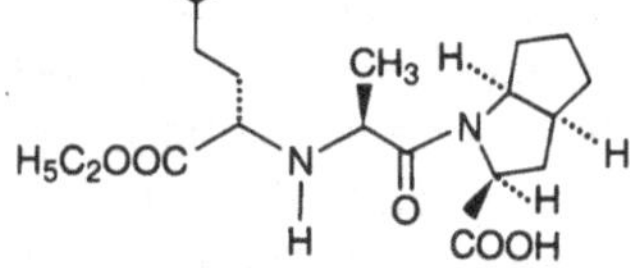

(79)

Today, captopril (Capoten®) ranks among the most frequently used drugs in the treatment of hypertension. Enalapril (Xanef®) has been commercially available since 1985 as a second ACE inhibitor. The discovery of captopril started an avalanche of research into the synthesis of angiotensin-converting enzyme inhibitors. Some new developments should be mentioned at this point:

1. Tripeptide analogs of enalapril *80* [179], e.g.

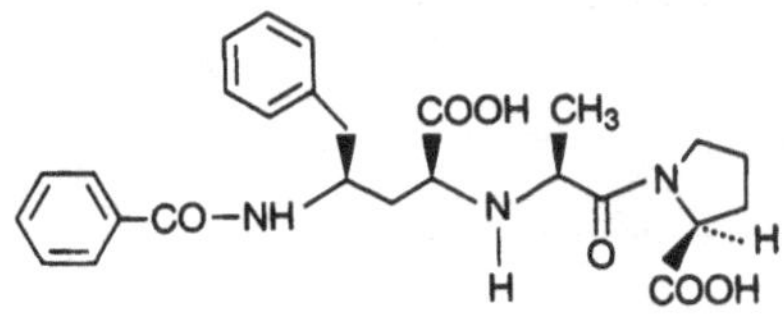

(80)

2. 1-Glutarylindoline-2-carboxylic acid derivatives *81* [180], e.g.

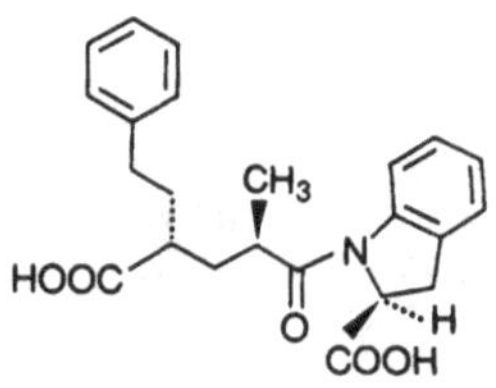

(81)

3. N-substituted γ-D-glutamyl-cis-perhydroindoline-2-(S)-carboxylic acid *82* [181], e.g.

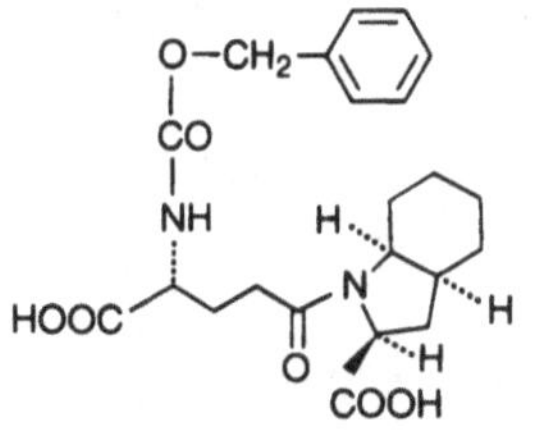

(82)

4. Cilazapril = (1S,9S)-9-[(S)-1-ethoxycarbonyl-3-phenylpropylamino]-octahydro-10-oxo-6-H-pyridazo-[1,2-a][1,2] diazepine-carboxylic acid *83* [182]:

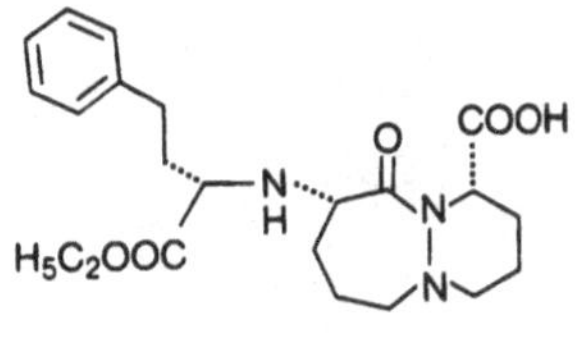

(83)

5.3 Renin Inhibitors

Another way of reducing the blood pressure via the renin-angiotensin system is to block the conversion of angiotensinogen into angiotensin I by inhibition of renin. Two different types of renin inhibitors are distinguished.

5.3.1 Substrate-Analogous Renin Inhibitors

As a highly specific acid peptidase, renin cleaves the decapeptide angiotensin I from the N-terminal end of the substrate angiotensinogen (Fig. 12).

ACE ↓ (Phe-His); Renin ↓ (Leu-Val)

Asp-Arg-Val-Tyr-Ile-His-Pro-Phe-His-Leu-Val-Ile-His-X-Glu-Protein

Angiotensin II (Asp … Phe)

Angiotensin I (Asp … Leu)

Angiotensinogen (Asp … Protein)

Fig. 12. Stepwise enzymatic cleavage of human angiotensinogen; X = peptide fragment of angiotensinogen

Derivatizations in the N-terminal sequence of human angiotensinogen led to weakly active renin inhibitors [183]. According to Szelke et al. [184], highly active substrate analogs with modified peptide linkage (—CH_2—NH—) were formed by reduction of the CO—NH bond in Leu-Leu or Phe-Phe (Table 2).

Table 2. Renin inhibitors with a modified peptide linkage; R = reduced peptide linkage

Substrate sequence 1 2 3 4 5 6 7 8 9 10 11 12 13 14 Asp Arg Val Tyr Ile His Pro Phe His Leu Leu Val Tyr Ser ...		Plasma, human IC_{50} [M] K_I
	R His Pro Phe His Leu-Leu Val Tyr	1×10^{-6}
H-77	R D-His Pro Phe His Leu-Leu Val Tyr	1×10^{-6}
	R His Pro Phe His Phe-Phe Val Tyr	8.2×10^{-7}
	R His Pro Phe His Leu-Val Ile His	1.9×10^{-7}
	R His Pro Phe His Leu-Val Ile Tyr	1×10^{-7}
	Pro His Pro Phe His Leu-Val Ile His	1.6×10^{-8}
H-142	R Pro His Pro Phe His Leu-Val Ile His Lys	1.0×10^{-8}
Pepstatin	Iva Val Val Sta-Ala-Sta	2.2×10^{-5}

The results of in vitro and human volunteer studies show that the best renin inhibitor to date is H-142.

5.3.2 Statine-Containing Renin Inhibitors

Pepstatin *84*, which was isolated from actinomycetes and which contains the unusual amino acid statine (Sta), inhibits renin and other acidic proteases.

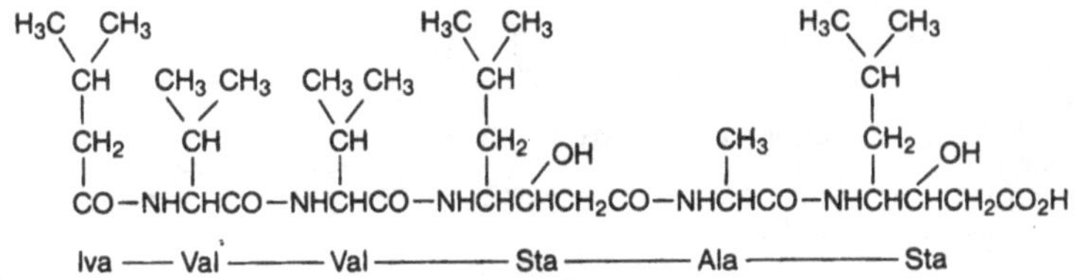

(84)

By exchanging Leu^{10} for Sta in the N-terminal end of angiotensinogen, active renin inhibitors have been produced (Table 3) [185–188]. It is assumed that statine corresponds to the tetrahedral intermediate during the enzymatic hydrolysis of the Leu^{10}-Val^{11} peptide linkage in angiotensinogen.

Table 3. Inhibition of porcine renin by statine-containing substrate analogs, (*K_M-value [183])

6 7 8 9 10 11 12 13	IC_{50} (M)
His–Pro–Phe–His–Leu–Leu–Val–Tyr	(5.5×10^{-5})*
His–Pro–Phe–His–Sta–Leu–Phe–NH_2	2.0×10^{-8}
Iva–His–Pro–Phe–His–Sta–Leu–Phe–NH_2	3.1×10^{-8}
Boc–His–Pro–Phe–His–Sta–Leu–Phe–NH_2	2.7×10^{-8}
Sta–Leu–Phe–NH_2	2.0×10^{-3}
His–Sta–Leu–Phe–NH_2	3.7×10^{-4}
Phe–His–Sta–Leu–Phe–NH_2	1.3×10^{-6}
Pro–Phe–His–Sta–Leu–Phe–NH_2	2.0×10^{-7}
Ibu–His–Pro–Phe–His–Sta–NH_2	2.9×10^{-5}
Ibu–His–Pro–Phe–His–Sta–Leu–NH_2	6.7×10^{-7}
Ibu–His–Pro–Phe–His–Sta–Leu–Phe–NH_2	4.3×10^{-8}
Ibu–His–Pro–Phe–His–Sta–Ala–Phe–NH_2	5.7×10^{-8}
Ibu–His–Pro–Phe–His–Sta–Val–Phe–NH_2	1.2×10^{-7}
Iva–His–Pro–Phe–His–Sta–Ile–Phe–NH_2	1.3×10^{-7}
Boc–His–Pro–Phe–His–Sta–Leu–Tyr–NH_2	2.6×10^{-8}
Boc–His–Pro–Phe–His–Sta–Leu–Phe–OCH_3	1.1×10^{-8}
Iva–His–Pro–Phe–His–Sta–Leu–Val–Phe–NH_2	4.6×10^{-5}

Iva = isovaleroyl Boc = *tert*-Butoxycarbonyl

Iva-His-Pro-Phe-His-Sta-Ile-Phe-NH_2 is so far the best statine-containing renin inhibitor.

According to Rich, new protease inhibitors were produced by replacing statine in pepstatin derivatives with (3S,4S)-4,8-diamino-3-hydroxyoctanecarboxylic acid (DAHOA) *85a* [189a)] or 4-amino-3-hydroxy-5-phenylpentanecarboxylic acid (AHPPA) *85b* [189b, c)].

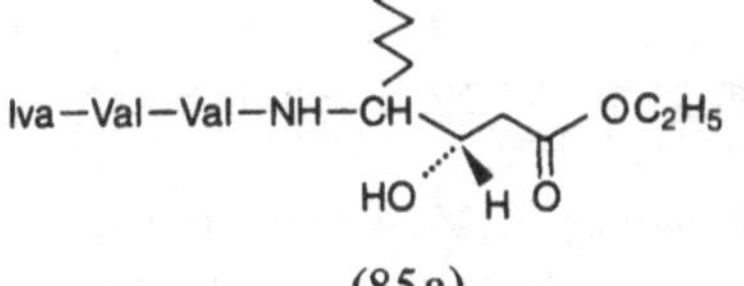

(85a)

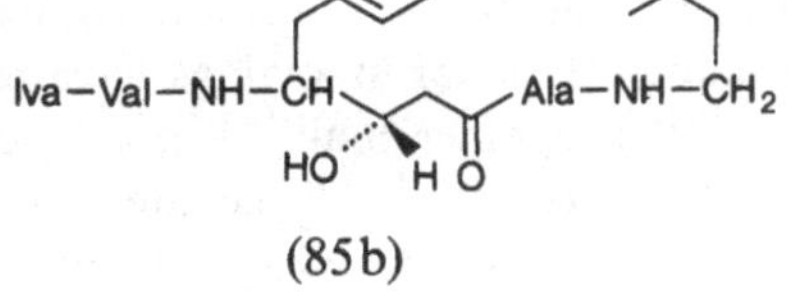

(85b)

L 364210 *86* and analogous highly active renin inhibitors, which have a longer duration of action and greater stability in vivo owing to their improved polarity and solubility, have been produced by Merck & Co./USA [190].

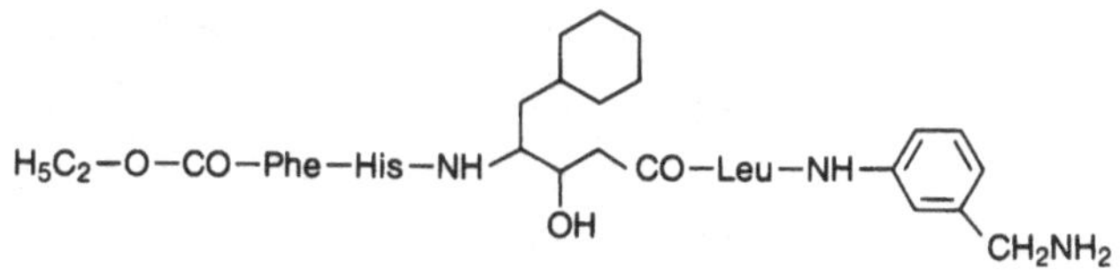

(86)

It is not only of scientific but also of great commercial interest to develop new drugs active on renin-induced hypertension.

6 Plasma Kinins

Plasma kinins are tissue hormones liberated from α-globulins of the blood plasma by kallikrein.

6.1 Bradykinin and Kallidin

Bradykinin *87* and kallidin (Lys-bradykinin-decapeptide), which are split off from the kininogen in the plasma by trypsin and kallikrein respectively (Fig. 10), hardly differ in their pharmacological activity. The most important effect of the kinins is a dilation of the peripheral vessels, which leads to an improved blood flow, in the kidneys for example, and therefore increases diuresis. By acting on the formation of angiotensin II, kinins can contribute to the regulation of blood pressure. Moreover, kinins cause a contraction of the bronchial muscle.

Arg-Pro-Pro-Gly-Phe-Ser-Pro-Phe-Arg (87)

Bradykinin was first synthesized by Boissonnas et al. [191], since when many research groups have reported on bradykinin analogs [89].

6.2 Tachykinins (Fig. 13)

The following active substances, known as tachykinins [192], which in contrast to the slow-acting kinins exert a rapid stimulating effect on the smooth muscle, are similar in their biological activities and their structure but differ in their origin. Eledoisin, which was discovered in the salivary glands of cephalopods from the Mediterranean by Erspamer in 1949, exerts an antihypertensive effect and acts as a spasmogen. A large number of eledoisin analogs have been prepared by Boissonnas et al. [193], Lübke et al. [194], Erspamer et al. [195], and Voelter et al. [196].

Physalaemin [197, 198], isolated by Erspamer in 1964 from the skin of American amphibians, has a strong vasodilating and antihypertensive action.

Uperolein, phyllomedusin, and kassinin, which Anastasi[199] isolated from frog skin, exhibit broad similarities with other tachykinins.

Pyr-Pro-Ser-Lys-Asp-Ala-Phe-Ile-Gly-Leu-Met-NH_2	Eledoisin
Pyr-Ala-Asp-Pro-Asn-Lys-Phe-Tyr-Gly-Leu-Met-NH_2	Physalaemin
Pyr-Ala-Asp-Pro-Lys-Thr-Phe-Tyr-Gly-Leu-Met-NH_2	[Lys^5,Thr^6]Physalaemin
Pyr-Pro-Asp-Pro-Asn-Ala-Phe-Tyr-Gly-Leu-Met-NH_2	Uperolein
Pyr — Asn-Pro-Asn-Arg-Phe-Ile-Gly-Leu-Met-NH_2	Phyllomedusin
Asp-Val-Pro-Lys-Ser-Asp-Gln-Phe-Val-Gly-Leu-Met-NH_2	Kassinin
Asp-Glu-Pro-Lys-Pro-Asp-Gln-Phe-Val-Gly-Leu-Met-NH_2	[Glu^2,Pro^5]Kassinin
Arg-Pro-Lys-Pro-Gln-Gln-Phe-Phe-Gly-Leu-Met-NH_2	Substance P
Arg-His-Lys-Thr-Asp-Ser-Phe-Val-Gly-Leu-Met-NH_2	Substance K

Fig. 13. Primary structures in the family of tachykinins; Pyr = pyroglutamic acid

Substances P and K can be considered as mammalian tachykinins in view of their unmistakable chemical relationship.

Caerulein *88*, which was isolated from skin extracts from the Australian tree frog *Hyla caerula*, has a longer-lasting antihypertensive activity than bradykinin or physalaemin. It causes a contraction of the gallbladder and bile ducts, and stimulates intestinal peristalsis. Caerulein analogs have been synthesized by Bernardi et al.[200, 201].

$$\text{pyroGlu-Gln-Asp-}\underset{\displaystyle SO_3H}{\text{Tyr}}\text{-Thr-Gly-Trp-Met-Asp-Phe-}NH_2 \qquad (88)$$

7 Atrial Natriuretic Peptides (ANP or ANF)

In 1983 de Bold et al.[202] first isolated an atrial peptide ANP-(6-33) from homogenates of rat atrial muscle and elucidated its structure (Fig. 14). In January 1984, K. Kangawa et al. prepared pure samples of the α-human atrial natriuretic peptide α-h-ANP-(6-33 Met^{17})[203] and of the hitherto longest-known natriuretic peptide containing 126 amino acids, γ-h-ANP[204], from human atrial tissue.

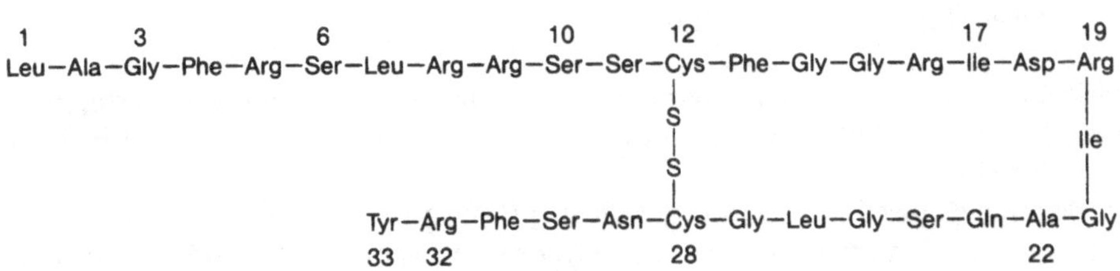

Fig. 14. Amino acid sequence of rat atrial natriuretic peptide

Seidah et al. [205] isolated ANF-(1-33) from rat atria and synthesized a shorter ANF-(8-33), which is identical with natural ANF-(3-33) in its biological activity.

Needleman et al. [206] isolated atriopeptin I [ANF-(10-30) or AP I] and atriopeptin II [ANF-(10-32) or AP II], also from rat atrium extracts.

S. P. Adams (Monsanto Compay) and P. Needleman (Washington University, St. Louis) synthesized the following atrial peptides by the solid-phase method [207] (Table 4).

Table 4. Biological activity of ANF analogs

ANF analogs	Relative activity in rabbit aorta assays
AP I	1.4
AP II	200
AP^{3-19}	0.35
Arg-Arg-AP II	300
Ser-Leu-Arg-Arg-AP II	150
$DesPhe^{22}$-AP II	7.0
$DesSer^{1}$-AP II	200
$DesSer^{1}$-$DesSer^{2}$-AP II	56
$DesSer^{21}$-AP I	1.0

Table 4 shows that arginine residues at the N-terminal end increase the bioactivity. Tyr^{33} is not absolutely necessary, but all the other amino acids at the C-terminal are (see AP I).

The ANF hormones, which derive from higher-molecular-weigth precursors (atriopeptigens), have diuretic properties, i.e. an administration of ANF in the rat increases diuresis and natriuresis (the release of Arg-vasopressin is inhibited) and at the same time the vessels are dilated, apparently by inhibition of catecholamines and angiotensin II. In addition, it has been shown that under volume loading the ANF peptides are released from the atria and develop their effects as hormones in renal, vascular, and other tissues. They can be considered as functional antagonists of the renin-angiotensin system.

The antihypertensive properties of ANF could really represent a new therapeutic starting point in combatting hypertensive disease in man.

8 Thyroid Hormones (Calcitonins)

Calcitonins (Fig. 15) with an N-terminal intrachain 23-membered disulfide ring were isolated independently by 4 different research groups, structurally clarified in 1968, and synthesized.

Calcitonin causes the deposition of calicum phosphate in the skeleton by stimulation of the bone-forming cells, and hence reduces the levels of calcium and phosphate in the blood (hypocalcemia). Completely synthetic calcitonin products such as Salmcalcitonin (Sandoz Ltd., Switzerland) [208], Elcitonin (Toyo Jozo Co., Japan) [209], and

Bovine H-Cys-Ser-Asn-Leu-Ser-Thr-Cys-Val-Leu-Ser-Ala-Tyr-Trp-Lys-Asp-Leu-

Salmon H-Cys-Ser-Asn-Leu-Ser-Thr-Cys-Val-Leu-Gly-Lys-Leu-Ser-Gln-Glu-Leu-

Human H-Cys-Gly-Asn-Leu-Ser-Thr-Cys-Met-Leu-Gly-Thr-Tyr-Thr-Gln-Asp-Phe-

Bovine Asn-Asn-Tyr-His-Arg-Phe-Ser-Gly-Met-Gly-Phe-Gly-Pro-Glu-Thr-Pro-NH_2

Salmon His-Lys-Leu-Gln-Thr-Tyr-Pro-Arg-Thr-Asn-Thr-Gly-Ser-Gly-Thr-Pro-NH_2

Human Asn-Lys-Phe-His-Thr-Phe-Pro-Gln-Thr-Ala-Ile-Gly-Val-Gly-Ala-Pro-NH_2

Fig. 15. Primary structures of calcitonin in various species

[16-alanine]-Salmcalcitonin (Armour Pharmaceutical Co.)[210], are available for therapeutic use (bone atrophy).

9 Peptide Antibiotics

Since the discovery of bacitracin *89* 35 years ago and of gramicidin *90* [211] in 1946, hundreds of peptide antibiotics have been synthesized.

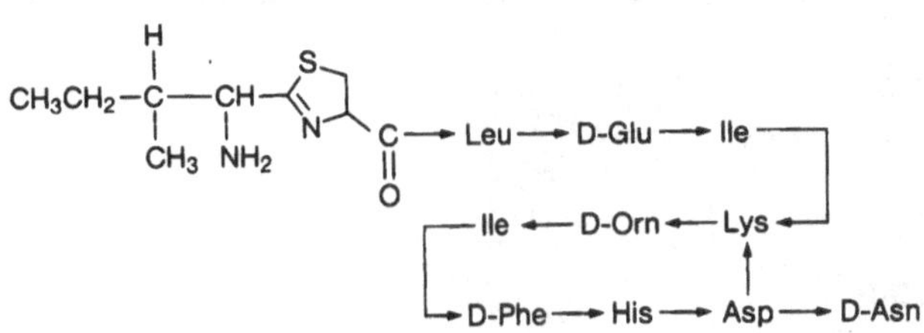

(89)

cyclo[–Val–Orn–Leu–D–Phe–Pro–Val–Orn–Leu–D–Phe–Pro–] (90)

These are active against gram-positive but not against gram-negative bacteria.

In comparison with the penicillin and cephalosporin derivatives, the peptide antibiotics are not numbered among the "major antibiotics". Their action mechanisms vary, e.g. inhibition of cell-wall synthesis, increased permeability of the cell wall, or influence on nucleic acid synthesis.

The presence of D-amino acids and other unusual non-proteinogenic amino acids is characteristic.

9.1 Monamycins

The monamycins *91*, a family of 15 hexapeptide members, which, as ionophores, induce the passage of ions through biological membranes, have hexahydropyridazine-1-carboxylic acid as their characteristic basic unit and exhibit antibacterial properties.

Hassall et al. [212] have synthesized cyclic hexadepsipeptides ($R^1 = R^2 = R^4 = H$, $R^3 = CH_3$), which form strong complexes with K^+, Rb^+, and Cs^+.

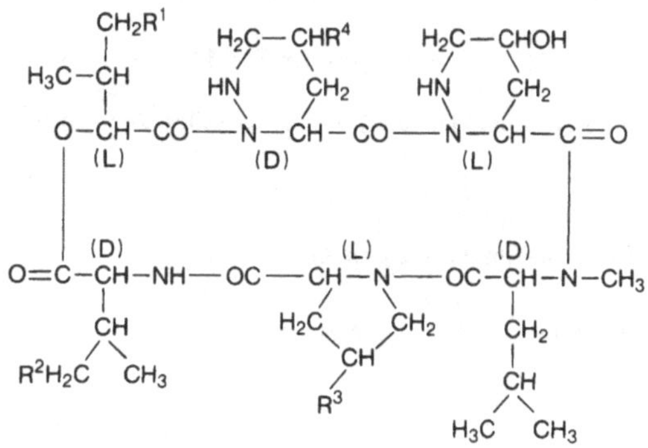

(91)

9.2 Phosphonopeptides [213–215]

Peptide antibiotics, for example (S)-alanyl-(R)-1-aminoethylphosphonic acid (alafosfalin) *92*, having an aminophosphonic acid at the C-terminal end of a peptide chain, have been synthesized by Hoffmann-La Roche AG/Switzerland [216a)] and Roche Products Ltd./U.K. [216b)]. Alafosfalin, which inhibits the biosynthesis of the bacterial cell wall, is effective against gram-positive and gram-negative microorganisms.

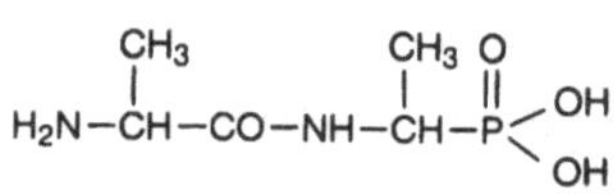

(92)

9.3 Actinomycins

Actinomycin D *93*, isolated from *Streptomyces antibioticus* in 1940, belongs to the class of chromopeptides and is characterized by its cytostatic growth inhibition in tumors and antibacterial action.

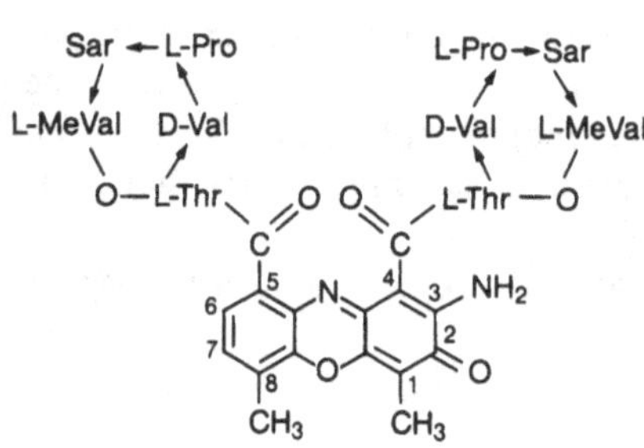

(93)

More than 30 natural actinomycins are now known and a variety of synthetic ones [89, 217], linked with 2 pentapeptide lactone rings via an aminophenoxazinone chromophore.

The use of the actinomycins is limited by their high toxicity.

9.4 Albomycins

Albomycins (desferriforms, Fig. 16), isolated from the strain *Streptomyces spec. WS 116*, are nucleoside peptides that exert antibiotic effects [218] and have iron-complexing properties.

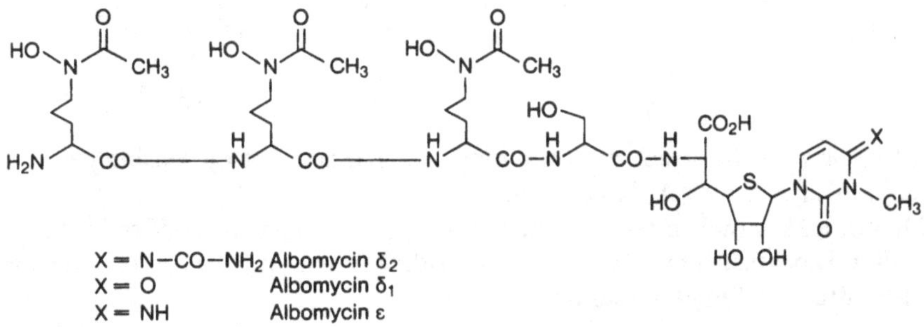

Fig. 16. Desferriforms of albomycins

9.5 Nisin

Nisin (Fig. 17), isolated from *Streptococcus lactis* culture broths in 1952, was structurally clarified only in 1970 by Gross et al. [219]. A partial sequence (ring A) of nisin was synthesized in 1983 by Shiba et al. [220].

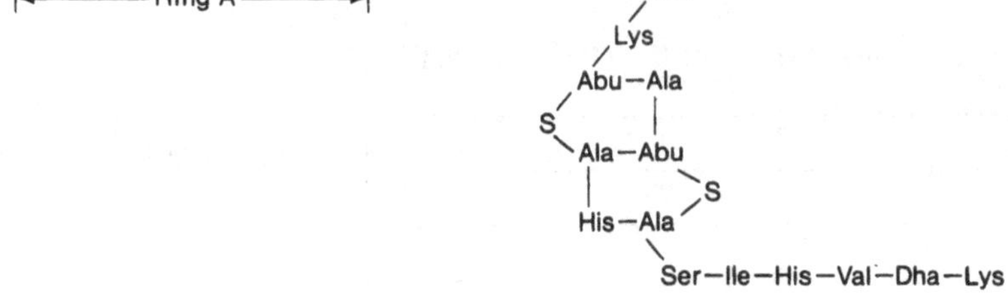

Fig. 17. Amino acid sequence of nisin; Abu = 2-aminobutyric acid

Nisin served to confirm that α,β-unsaturated amino acids occur naturally (Dha = dehydroalanine, Dhb = α-aminodehydrobutyric acid).

9.6 Nikkomycins

Nikkomycins Z *94* and X *95*, isolated as nucleoside antibiotics from *Streptomyces tendae* [221], inhibit chitin biosynthesis and have fungicidal and insecticidal properties.

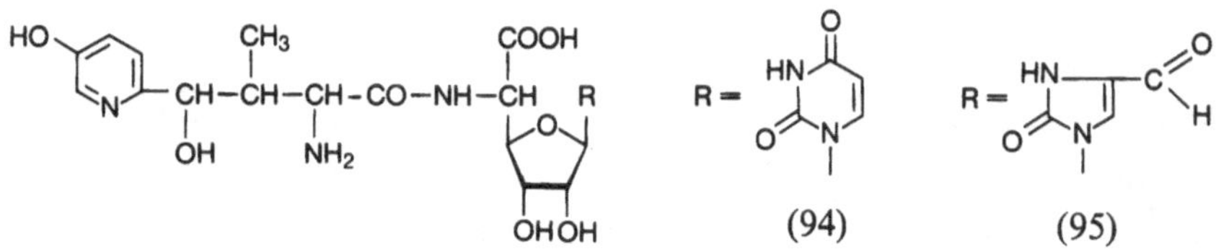

(94) (95)

Several reports on the structure elucidation and syntheses of nucleoside peptides have been published by W. A. König et al. [222].

In addition, nikkomycin B *96* [223], which has a p-hydroxyphenyl residue in the N-terminal amino acid instead of the 3-hydroxypyridine system of *95*, was isolated from the culture filtrate of *Streptomyces tendae*.

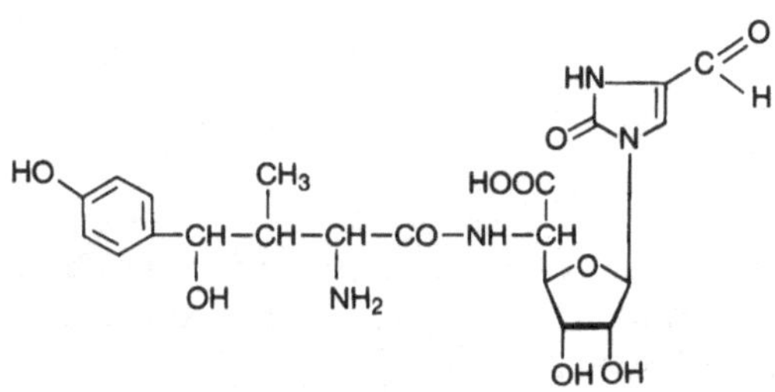

(96)

9.7 Cecropin A

Cecropins [224] are produced in insects on account of the lack of lymphocytes and immunoglobulins by a humoral immune reaction, and have a broad spectrum of antibacterial activity.

Cecropin A analogs (Fig. 18) have been synthesized by Andreu et al. [225] by the solid-phase method.

Lys^{1} Trp Lys Leu Phe Lys Lys Ile Glu Lys Val^{11}
——— Glu ———————————————————
——————— Pro ———————————————
——————————————— Glu ———————
——————————————— Leu ———————
————————————————————— Pro ———

Gly^{12} Gln Asn Ile Arg Asp Gly Ile Ile Lys Ala^{22}
Gly^{23} Pro Ala Val Ala Val Val Gly Gln Ala Thr^{33}
Gln^{34} Ile Ala Lys NH_2

Fig. 18. Amino acid sequences of cecropin A analogs

9.8 Alamethicins

In 1970 Payne et al. [226] elucidated the lipophilic α-helical structure of the eicosapeptide alamethicin *97* (eight α-aminobutyric acid residues = Aib, L-phenylalaninol = Phl), isolated from the culture liquid of the fungus *Trichoderma viride*. The sequence of the amphiphilic peptide antibiotic was confirmed in 1985 by Jung et al. [227] by total syntheses.

1 2 3 4 5 6 7 8 9 10 11 12 13 14 15 16 17 18 19 20
Ac-Aib-Pro-Aib-Ala-Aib-Ala-Gln-Aib-Val-Aib-Gly-Leu-Aib-Pro-Val-Aib-Aib-Glu-Gln-Phl (97)

Alamethicin exerts bacteriostatic, fungicidal, cytostatic, and hemolytic effects. The most important property of the alamethicins is the formation of potential-dependent ion-conducting pores in lipid membranes as a model for the conduction of nerve impulses [227 b)].

10 Peptide Leukotrienes [228]

Leukotriene LTC_4 *99* is formed enzymatically from leukotriene LTA_4 *98*, which is formed from arachidonic acid by means of lipoxygenase, by nuclephilic attack of the thiol group in glutathione (γ-glutamylcysteinylglycine).

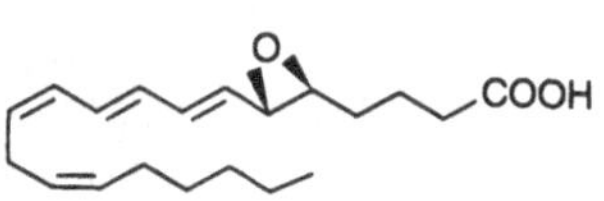

(98)

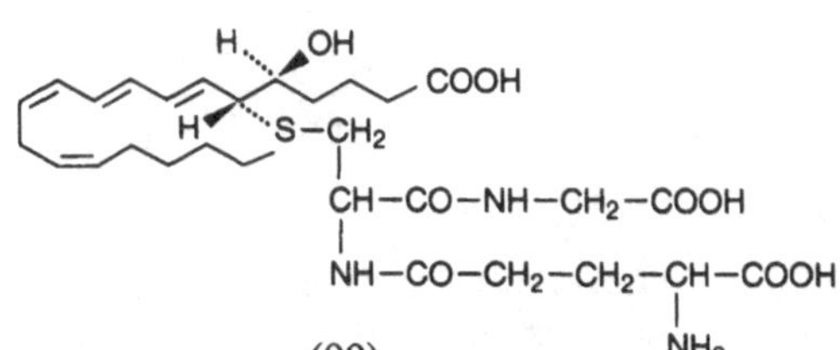

(99)

LTC_4 (slow-reacting substance = SRS), synthesized by Corey et al. [229], is converted into the biologically more active S-cysteine-glycylleukotriene LTD_4 *100* or S-cysteinyl-leukotriene LTE_4 *101* under the influence of γ-glutamyltranspeptidase (GGTP).

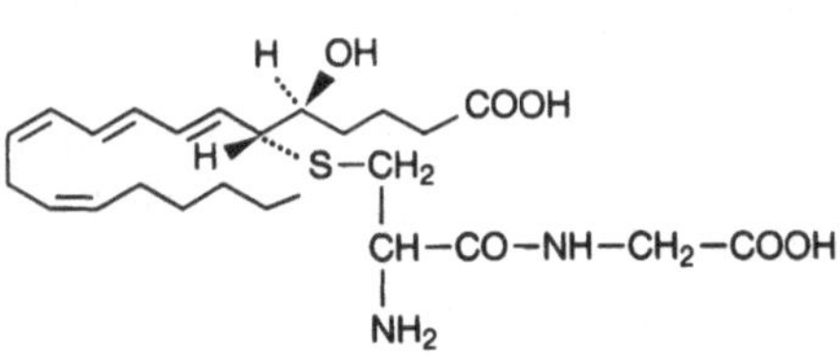

(100)

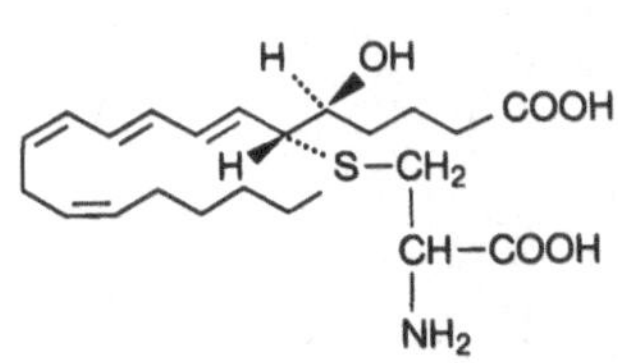

(101)

The peptide leukotrienes *99*, *100*, and *101* cause contraction of the bronchial smooth muscle and probably play an important role as mediators in allergic reactions (e.g. asthma) and inflammations. Antagonistic blockade of the leukotriene action of *99*, *100*, and *101*, by analogy with histamine H_1-receptor antagonists, would therefore be an important principle in the treatment of allergic symptoms. A number of selective peptide leukotriene antagonists have in fact been synthesized by Smith Kline &

French Lab. [230a)] and Schering-Plough Co./USA [230b)], e.g. 4-R-hydroxy-5-S-cysteinyl-glycyl-6Z-nonadecenoic acid *102*.

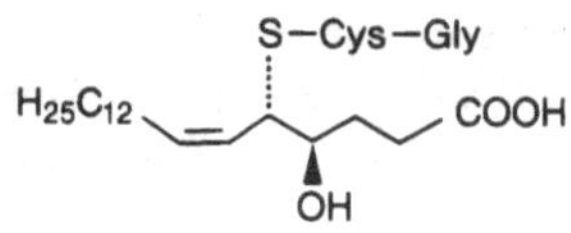

(102)

11 Peptide Insecticides and Herbicides

The cyclic depsipeptides destruxin C and D *103* and bassianolid *104* [231)], which contain α-hydroxyisovaleric acid, N-methylvaline, or N-methylleucine, act as insecticides.

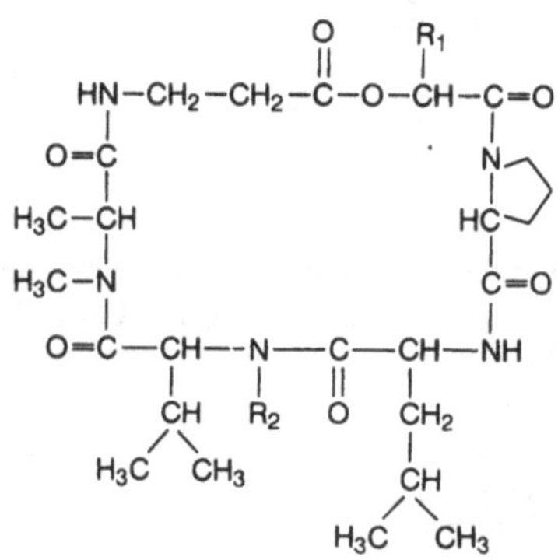

(103)

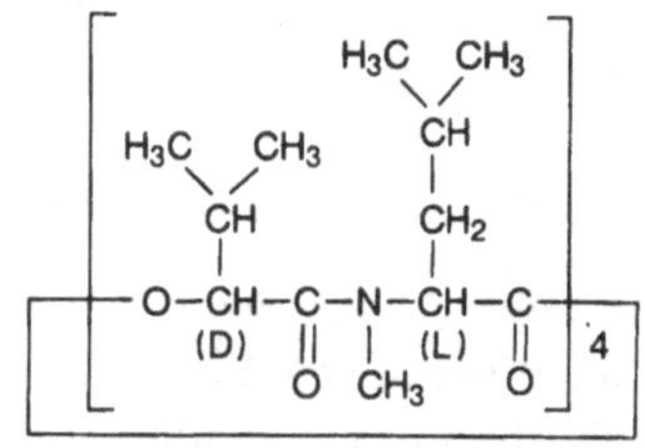

(104)

Destruxin	R_1	R_2
C	$HO-CH_2-CH(CH_3)-CH_2-$	H_3C-
D	$H_3C-CH(COOH)-CH_2-$	H_3C-

Phosphinotricylalanylalanine *105* (bialaphos), isolated by Zähner et al. [215, 232)] from the culture filtrates from *Streptomyces* species, has a strong herbicidal action.

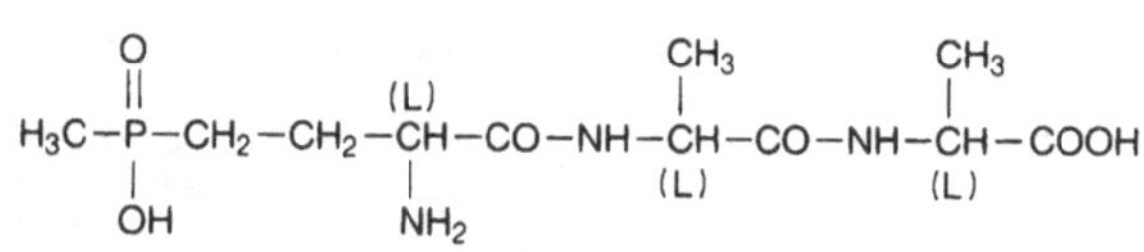

(105)

12 Peptide Toxins

The phallotoxins *106*, e.g. phalloidin, and the amatoxins *107*, e.g. α-amanitin, produced by *Amanita phalloides* or death cup, are among the best-known peptide poisons [233)].

The amatoxins are cyclic octapeptides composed only of L-amino acids and containing a sulfoxide group instead of the thioether bridge in phallotoxin. Over 90% of the fatal cases of mushroom poisoning can be traced back to the amatoxins. Wieland et al. [234)] have shown that, in addition to the toxins, the death cup contains a low concentration of an antitoxic cyclic decapeptide antamanide *108*.

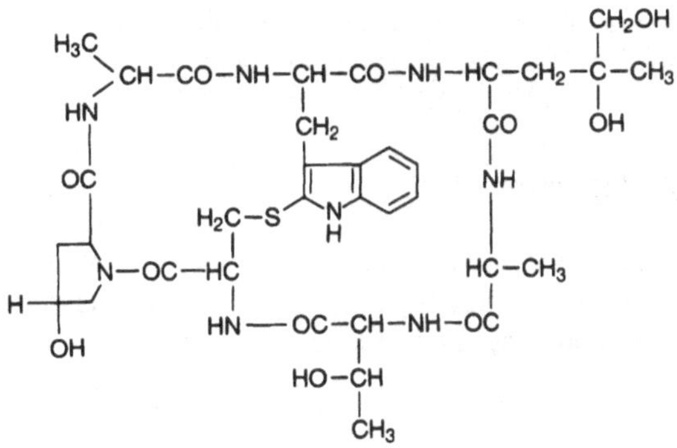

(106)

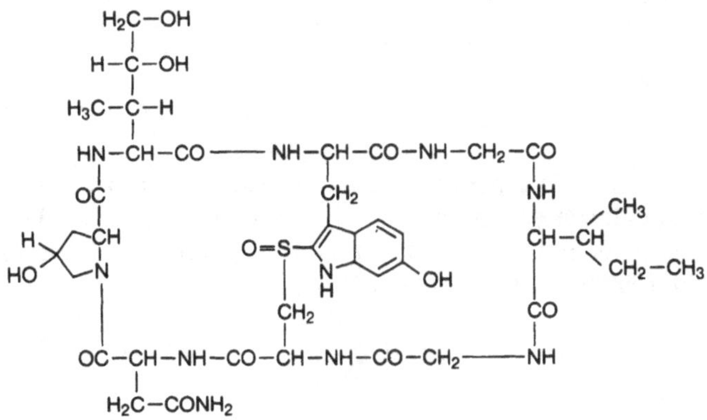

(107)

cyclo-(-Pro-Phe-Phe-Val-Pro-Pro-Ala-Phe-Phe-Pro-) (108)

The cyclopentapeptide malformin *109*, a metabolite of *Aspergillus niger*, was structurally elucidated by Bodanszky [235a)] and synthesized in 1973. This has antibiotic and antimycotic properties and causes malformation in mice and in higher plants.

→D-Cys-D-Cys-L-Val-D-Leu-L-Ile—
1 2 3 4 5 (109)

The sequence isomer of the natural product [Ile^3, Val^5]-malformin *110* was synthesized as allomalformin by Bodanszky et al. [235b)] in 1982.

→D-Cys-D-Cys-L-Ile-D-Leu-L-Val— (110)

The structure of the toxic octapeptide lophyrotomin *111* (LD_{100} = 2 mg/kg), isolated from Australian sawfly larvae (*Lophyrotoma interrupta*), was established by D. H. Williams et al. [236)] in 1983.

C_6H_5CO-(D)-Ala-(D)-Phe-(L)-Val-(L)-Ile-(D)-Asp-(L)-Asp-(D)-Glu-(L)-Gln (111)

Lophyrotomin leads to fatal intoxications in cattle and sheep, with muscle twitching, refusal of food, and acute liver failure.

13 Sweet and Bitter Peptides

During the synthesis of a gastrin tetrapeptide Schlatter et al. [237] made the incidental but extremely interesting discovery that the dipeptide aspartylphenylalanine methyl ester *112* (aspartame) is 100–200 times as sweet as sucrose.

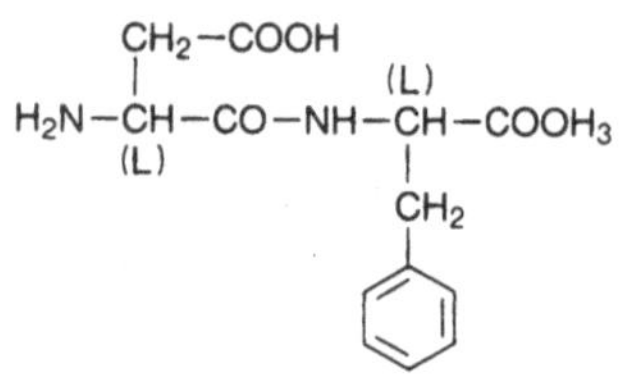

(112)

Aspartame is marketed very successfully as a sweetener in the USA.

The 1,1-diaminoalkane derivatives such as *113*, developed as a new class of sweet peptides by Goodman et al. [238] on the basis of the retro-inverso peptide modification [20], are 800–1000 times as sweet as sucrose.

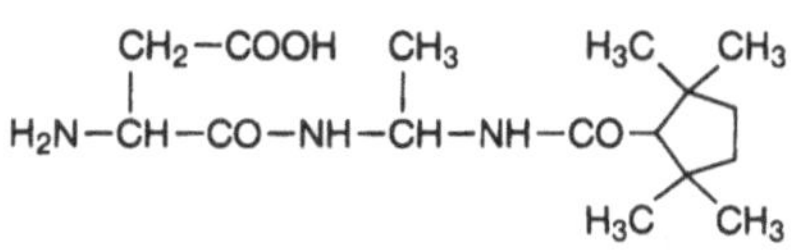

(113)

The bitter peptide BPI a *114*, isolated by Okai et al. [239] from casein hydrolysates, and delicious tasting peptides from fish proteins, will undoubtedly achieve practical importance in the food industry.

Arg-Gly-Pro-Pro-Ile-Val (114)

14 Final Remarks

The purpose of this review was to show the variety of peptides synthesized or isolated from natural products in the last 15 years and to classify the biologically active peptides into the various categories, according to how and where they are formed, their transport, and their general cellular activity. The number of biologically active peptides has risen sharply in the last few years owing to the improvement in the preparative methods in conventional peptide synthesis — at present about 130 different coupling methods are known — and in the solid-phase peptide synthesis.

New interesting peptide hormones have also been found in human, animal, and vegetable organs owing to the improvements in methods of analysis and separation. For example, the development of radioimmunoassay first paved the way for the investigation of neuropeptides. Thus, R. Guillemin and A. V. Schally were able to establish the structure and to synthesize the first hypothalamic hormones. The discovery of a stereospecific opiate receptor in the nervous system initiated an intensive search for endogenous substrates for this receptor (endorphins, dynorphins). We are probably only at the beginning of the development of neuropeptides (e.g. CNS-active peptides), whose end is still not in sight.

For some years DNA-recombination has also been available for peptide synthesis. Genetic engineering will enable peptides of higher molecular weight, and those hardly accessible up to now, to be obtained in larger amounts.

The predominantly clinical use of peptide pharmaceuticals and their applications in diagnostics have so far kept their market potential within narrow limits. Since the discovery of angiotensin II antagonists (e.g. saralasin) and ACE inhibitors (e.g. captopril), however, peptide chemistry has gained in importance within the context of drug research. Broader introduction of active peptide substances in pharmacotherapy will have to await the development of peptide drugs suitable for administration by the oral route.

15 Acknowledgements

Special thanks are due to Ilona Gabriele Deus for typing the manuscript and to Herr E. Müller, IN-AN Engineering Graphics Department of BAYER AG, Leverkusen, for the drawings.

16 References

1. Wünsch, E.: Synthese von Peptiden, Houben-Weyl, Bd. 15/1 und 2, Georg Thieme Verlag, Stuttgart 1974
2. a) Schröder, E., Lübke, K.: The Peptides I, Academic Press, New York 1965
   b) Schröder, E., Lübke, K., Kloss, G.: Chemie und Biochemie der Aminosäuren, Peptide und Proteine I und II, Georg Thieme Verlag, Stuttgart 1975
3. Gross, E., Meienhofer, J.: The Peptides, Vol. 1, 2, 3, Academic Press, New York 1979, 1980, 1981
4. Finn, F. M., Hofmann, K.: The Synthesis of Peptides by Solution Methods with Emphasis on Peptide Hormones, in: The Proteins, Vol. II (ed. Neurath, H., Hill, R. L.), p. 105, Academic Press, New York 1976
5. Jakubke, H.-D., Jeschkeit, H.: Aminosäuren, Peptide, Proteine, Verlag Chemie, Weinheim 1982
6. Bláha, K.: Linear Peptides: Synthetic Methods; Wade, R.: Linear Peptides: Synthesis and Structure — Activity Relationships; Wieland, T., Birr, C.: Homodetic Cyclic Peptides, in: Amino Acids, Peptides and Related Compounds, International Review of Science, Vol. 6 (ed. Rydon, H. N.), p. 73, 97, 183, Butterworths, London 1976
7. Stammer, C. H.: Procedures for Peptide Synthesis; Wade, R.: New Syntheses of Naturally Occurring Peptides and Analogues, in: Amino Acids, Peptides and Related Compounds, MTP International Review of Science, Vol. 6 (ed. Hey, D. H., John, D. I.) p. 135, 161, Butterworths, London 1976
8. a) Bodanszky, M.: Principles of Peptide Synthesis, Springer Verlag, Berlin 1984
   b) Bodanzsky, M., Bodanzsky, A.: The Practice of Peptide Synthesis, Springer Verlag, Berlin 1982
9. a) Erickson, B. W., Merrifield, R. B.: Solid-Phase Peptide Synthesis, in The Proteins, Vol. II (ed. Neurath, H., Hill, R. L.), p. 255, Academic Press, New York 1976
   b) Barany, G., Merrifield, R. B.: Solid-Phase Peptide Synthesis, in: The Peptides, Vol. 2 (ed. Gross, E., Meienhofer, J.) p. 1, Academic Press, New York 1980

10. Birr, C.: Aspects of the Merrifield Peptide Synthesis, Springer Verlag, Berlin 1978
11. a) Pillai, V. N. R., Mutter, M.: New Perspectives in Polymer-Supported Peptide Synthesis, in: Topics in Current Chemistry (ed. Boschke, F. L.), p. 119, Springer Verlag, Berlin 1982
b) Atherton, E., Logan, C. J., Sheppard, R. C.: J. Chem. Soc. Perkin Trans. I *1981*, 538
c) Sheppard, R. C.: Chemistry in Britain *19*, 402 (1983)
12. a) Mutter, M., Hagenmaier, H., Bayer, E.: Angew. Chemie *83*, 883 (1971)
b) Mutter, M., Bayer, E.: The Liquid-Phase Method for Peptide Synthesis, in: The Peptides, Vol. 2 (ed. Gross, E., Meienhofer, J.), p. 285, Academic Press, New York 1980
13. Jakubke, H.-D., Kuhl, P., Könnecke, A.: Angew. Chem. *97*, 79 (1985)
14. Kuhl, P., Zapevalova, N. P., Könnecke, A., Jakubke, H.-D.: Monatsh. Chem. *114*, 343 (1983)
15. Kullmann, W.: J. Org. Chem. *47*, 5300 (1982)
16. Chaiken, I. M., Komoriya, A., Ohno, M., Widmer, F.: Appl. Biochem. Biotechnol. *7*, 385 (1982)
17. Davies, J. E., Gassen, H. G.: Angew. Chem. *95*, 26 (1983)
18. Wengenmayer, F.: Angew. Chem. *95*, 874 (1983)
19. Spatola, A. F.: Peptide Backbone Modifications, in: Chemistry and Biochemistry of Amino Acids, Peptides and Proteins, Vol. 7 (ed. Weinstein, B.), p. 267, Dekker, New York 1983
20. Goodmann, M., Chorev, M.: The Synthesis and Conformational Analysis of Retro-Inverso-Analogues of Biologically Active Molecules, in: Perspectives in Peptide Chemistry (ed. Eberle, A., Geiger, R., Wieland, T.), p. 283, S. Karger, Basel 1981
21. Schröder, E., Rufer, C., Schmiechen, R.: Aminosäure-, Peptid- und Proteohormone, in: Pharmazeutische Chemie, p. 468, Georg Thieme Verlag, Stuttgart 1982
22. Greven, H. M., de Wied, D.: Neuropeptides and Behaviour, in: Perspectives in Peptide Chemistry (ed. Eberle, A., Geiger, R., Wieland, T.), p. 356, S. Karger, Basel 1981
23. Tregear, G. W., Niall, H. D., Potts, J. T., Leemann, S. E., Chang, M. M.: Nature, New Biol. *232*, 87 (1971)
24. Niedrich, H., Oehme, P., Mehlis, B., Bienert, M., Bergmann, J., Berger, H.: Substance P-Structural Features and Regulatory Functions, in: Perspectives in Peptide Chemistry (ed. Eberle, A., Geiger, R., Wieland, T.), p. 344, S. Karger, Basel 1981
25. Caranikas, S., Mizrahi, J., Escher, E., Regoli, D.: J. Med. Chem. *25*, 1313 (1982)
26. Chorev, M., Rubini, E., Gilon, C., Wormser, U., Selinger, Z.: J. Med. Chem. *26*, 129 (1983)
27. Pessi, A., Pinori, M., Verdini, A. S., Viscomi, G. C.: J. Chem. Soc. Commun. *1983*, 195
28. Munekata, E., Okada, M., Kimura, S., Sugita, Y., Kanazawa, I., Matsudo, T., Otsuka, M.: Chem. Lett. *1984*, 1013
29. a) Carraway, R., Leeman, S. E.: J. Biol. Chem. *248*, 6854 (1973)
b) Carraway, R., Leeman, S. E.: ibid. *250*, 1907 (1975)
30. St-Pierre, S., Lalonde, J.-M., Gendreau, M., Quirion, R., Regoli, D., Rioux, F.: J. Med. Chem. *24*, 370 (1981)
31. Hughes, J., Smith, T. W., Kosterlitz, H. W., Fothergill, L. A., Morgan, B. A., Morris, H. R.: Natur *258*, 577 (1975)
32. Blake, J., Tseng, L.-F., Chang, W.-C., Li, C. H.: Int. J. Peptide Protein Res. *11*, 323 (1978)
33. Yamashiro, D., Westphal, M., Holy, K., Li, C. H.: Int. J. Peptide Protein Res. *21*, 389 (1983)
34. Shimohiagashi, Y., Stammer, C. H.: J. Chem. Soc. Perkin Trans. I *1983*, 803
35. Deeks, T., Crooks, P. A., Waigh, R. D.: J. Med. Chem. *26*, 762 (1983)
36. Fauchère, J.-L., Pfenninger, S., Do, K. Q., Lemieux, C., Schiller, P. W.: Helv. Chim. Acta *66*, 1053 (1983)
37. Shimohiagashi, Y., Dunning, jr. J. W., Kolar, A. J., Stammer, C. H.: Int. J. Peptide Protein Res. *21*, 202 (1983)
38. Pless, J., Bauer, W., Cardinaux, F., Closse, A., Hauser, D., Huguenin, R., Roemer, D., Buescher, H.-H., Hill, R. C.: Helv. Chim. Acta *62*, 398 (1979)
39. Peterman, C., Fauchère, J.-L.: Helv. Chim. Acta *66*, 1513 (1983)
40. Richman, S. J., Goodman, M., Naguyen, T. M.-D., Schiller, P. W.: Int. J. Peptide Protein Res. *25*, 648 (1985)
41. Gesellchen, P. D., Shuman, R. T.: Brit.-Pat. 2065 132 (1980), E. Lilly
42. Shuman, R. T.: Brit.-Pat. 2065 135 (1981), E. Lilly
43. Mazur, R. H., Tyner, D. A., Hallinan, E. A.: EP 0 081 838 (1982), G. D. Searle & Co.
44. Morley, J. S.: EP 0 083 849 (1982), ICI Ltd.

45. a) Wilkinson, S., Wrigglesworth, R., Hardy, G. W.: EP 0 075 334 (1982), The Wellcome Foundation Ltd.
b) Schwartz, J.-C., Costentin, J., Lecomte, J.-M.: Trends in Pharmacological Sciences *6* (12), 472 (1985)
46. a) Llorens, C., Gilles, G., Swerts, J.-P., Perdrisot, R., Fournie-Zaluski, M.-C., Schwartz, J.-C., Roques, B. P.: Biochem. Biophys. Res. Commun. *96*, 1710 (1980)
b) Evans, D. A., Mathre, D. J.: J. Org. Chem. *50*, 1830 (1985)
47. Benovitz, D. E., Spatola, A. F.: Peptides *6*, 648 (1985)
48. Drugs of the Future *7*, 803 (1982)
49. Chang, K.-J., Killian, A., Hazum, E., Cuatrecasas, P.: Science *212*, 75 (1981)
50. Goldstein, A., Tachibana, S., Lowney, L. I., Hunkapiller, M., Hood, L.: Proc. Natl. Acad. Sci. USA *76*, 6666 (1979)
51. Li, C. H., Yamashiro, D., Ferrara, P., Tseng, L.-F., Way, E. L.: Int. J. Peptide Protein Res. *21*, 331 (1983)
52. Montecucchi, P. C., DeCastiglione, R., Erspamer, V.: Int. J. Peptide Protein Res. *17*, 316 (1981)
53. Yamashiro, D., Nicolas, P., Li, C. H.: Int. J. Peptide Protein Res. *21*, 219 (1983)
54. Salvadori, S., Marastoni, M., Balboni, G., Sarto, G. P., Tomatis, R.: J. Med. Chem. *28*, 769 (1985)
55. Takagi, H., Shiomi, H., Ueda, H., Amano, H.: Nature *282*, 410 (1979)
56. Takagi, H., Shiomi, H., Fukui, K., Hayashi, K., Kiso, Y., Kitagawa, K.: Life Sci. *31*, 1733 (1982)
57. a) Kitagawa, K., Kawai, N., Kiso, Y., Akita, T., Fukui, K., Amano, H., Takagi, H.: Chem. Pharm. Bull. *31*, 2349 (1983)
b) Kitagawa, K., Kawai, N., Kiyama, S., Akita, T., Kiso, Y., Ueda, H., Ming, G., Takagi, H.: Chem. Pharm. Bull. *33*, 377 (1985)
58. Schonenberger, G. A., Monnier, M.: Proc. Natl. Acad. Sci. USA *74*, 1282 (1977)
59. Mikhaleva, I., Sargsyan, A., Ivanov, V.: Proc. of the 17th Eur. Peptide Symp., p. 563, Prague 1982
60. Schmidt, G.: Pharmazie in unserer Zeit *1*, 181 (1972)
61. Bøler, J., Enzmann, F., Folkers, K., Bowers, C. Y., Schally, A. V.: Biochem. Biophys. Res. Commun. *37*, 705 (1969)
62. Burgus, R., Dunn, T. F., Desiderio, D., Ward, D., Vale, W., Guillemin, R.: Nature *226*, 321 (1970)
63. Rivier, J., Vale, W., Monahan, M., Ling, N., Burgus, R.: J. Med. Chem. *15*, 479 (1972)
64. a) Wissmann, H., König, W., Geiger, R.: DE 2 033 600 (1972), Hoechst AG
b) Folkers, K., Enzmann, F. H.: US-Pat. 3 746 697 (1973) Austin, USA
65. Fujino, M., Nishimura, O., Nagawa, Y., Fukuda, N.: DOS 2 712 086 (1977), Takeda
66. Drugs of the Future *7*, 167 (1982)
67. Schally, A. W., Arimura, A., Baba, Y., Nair, R. M. G., Matsuo, H., Redding, T. W., Debeljuk, L., White, W. F.: Biochem. Biophys. Res. Commun. *43*, 393 (1971)
68. a) König, W., Geiger, R., Sandow, J. K.: EP 0 041 243 (1980), Hoechst AG
b) König, W., Teetz, V., Jäger, G., Geiger, R.: DE 2 905 502 (1980), Hoechst AG
69. König, W., Geiger, R., Sandow, J.: DOS 2 438 350 (1974), Hoechst AG
70. a) Karcher, H. L.: Selecta *43*, 3548 (1984)
b) Drugs of the Future 7, 565 (1982) ibid. *4*, 173 (1979)
71. a) Coy, D. H., Vilchez-Martinez, J. A., Coy, E. J., Schally, A. V.: J. Med. Chem. *19*, 423 (1976)
b) Gendrich, R. L., Rippel, R. H., Seely, J. H.: US-Pat. 4 005 063 (1977), Abbott Lab., USA
c) Drugs of Today, Vol. 21, 393 (1985)
72. Nestor, J. J., Jones, G. H., Vickery, B. H.: DE 3 021 736 (1980), Syntex, USA
73. Nutt, R. F.: EP 0 070 021 (1982), Merck, USA
74. Nutt, R. F., Veber, D. F., Curley, P. E., Saperstein, R., Hirschmann, R.: Int. J. Peptide Protein Res. *21*, 66 (1983)
75. a) Veber, D. F., Holly, F. W., Strachan, R. G., Paleveda, W. J., Nutt, R. F., Hirschmann, R.: US-Pat. 4 161 521 (1979), Merck, USA
b) Sarantakis, D.: US-Pat. 4 093 609 (1978), American Home
76. Kamber, B.: EP 0 083 305 (1982), Ciba-Geigy
77. Drugs of the Future *9*, 342 (1984)
78. Drugs of the Future *6*, 491 (1981)
79. a) Veber, D. F., Nutt, R. F.: US-Pat. 4 191 754 (1980), Merck, USA

b) Freidinger, R. M., Veber, D. F.: J. Am. Chem. Soc. *101*, 6129 (1979)
c) Freidinger, R. M., Colton, C. D., Perlow, D. S., Whitter, W. L., Paleveda, W. J., Veber, D. F.: Proc. of the 8th Am. Peptide Symp., p. 349, Tucson 1983

80. a) Wynants, C., van Binst, G., Loosli, H. R.: Int. J. Peptide Protein Res. *25*, 608 (1985)
b) Lamberts, S. W. J., Uitterlinden, P., Verschoor, L., van Dongen, K. J., del Pozo, E.: New Engl. J. Med. *313*, 1576 (1985)
81. Vale, W., Spies, J., Rivier, C., Rivier, J.: Science *213*, 1394 (1981)
82. Morell, J. L., Brown, J. H.: Int. J. Peptide Protein Res. *26*, 49 (1985)
83. Schally, A. V., Redding, T. W., Chang, R. C. C., Arimura, A., Huang, W. Y., Coy, D. H., Meyers, C. A., Pedroza, E., Kastin, A. J., Turkelson, C., in: Polypeptide Hormones (ed. Beers, R. F., Bassett, E. G.), p. 169, Ravens Press, New York 1980
84. Guillemin, R., Brazeau, P., Böhlen, P., Esch, F., Ling, N., Wehrenberg, W. B.: Science *218*, 585 (1982)
85. Rivier, J., Spiess, J., Thorner, M., Vale, W.: Nature *300*, 276 (1982)
86. a) Fujii, N., Shimokura, M., Nomizu, M., Yajima, H., Shono, F., Tsuda, M., Yoshitake, A.: Chem. Pharm. Bull. *32*, 520 (1984)
b) Fujii, N., Lee, W., Shimoruka, M., Yajima, H.: ibid. *32*, 739 (1984)
87. a) Li, C. H., Yamashiro, D.: J. Am. Chem. Soc. *92*, 7608 (1970)
b) Blake, J., Li, C. H.: Int. J. Peptide Protein Res. *7*, 495 (1975)
c) Blake, J., Li, C. H.: ibid. *11*, 323 (1978)
88. Kappeler, H., Schwyzer, R.: Helv. Chim. Acta *44*, 1136 (1961)
89. Schröder, E., Lübke, K.: The Peptides II, Academic Press, New York 1966
90. Schwyzer, R., Sieber, P.: Helv. Chim. Acta *49*, 134 (1963)
91. a) du Vigneaud, V., Ressler, C., Swan, J. M., Roberts, C. W., Katsoyannis, P. G., Gordon, S.: J. Am. Chem. Soc. *75*, 4879 (1953)
b) Katsoyannis, P. G., Gish, D. T., du Vigneaud, V.: ibid. *79*, 4516 (1957)
92. Zaoral, M., Vávra, I., Machova, A., Sorm, F.: DOS 1 643 273 (1967), Czeskoslovenská akademie Prag
93. Vinick, F. J., Jung, S.: J. Org. Chem. *48*, 392 (1983)
94. Josefsson, L.: Int. J. Peptide Protein Res. *21*, 459 (1983)
95. Ferrero, P., Guidotti, A., Conti-Tronconi, B., Costa, E., Neuropharmacology *23*, 1359 (1984)
96. Jorpes, J. E., Mutt, V.: Acta Chem. Scand. *15*, 1790 (1961)
97. Bodanszky, M., Ondetti, M. A., Levine, S. D., Williams, N. J.: J. Am. Chem. Soc. *89*, 6753 (1967)
98. Ondetti, M. A., Narayanan, V. L., von Saltza, M., Sheehan, J. T., Sabo, E. F., Bodanszky, M.: J. Am. Chem. Soc. *90*, 4711 (1968)
99. Wünsch, E., Wendlberger, G.: Chem. Ber. *105*, 2508 (1972)
100. Uchiyama, M., Sato, T., Yoshino, H., Tsuchiya, Y., Tsuda, T., Konishi, M., Tsujii, M., Hisatake, Y., Koiwa, A.: Chem. Pharm. Bull. *33*, 1990 (1985)
101. Staub, A., Sinn, L. G., Behrens, O. K.: Science *117*, 628 (1953)
102. Bromer, W. W., Sinn, L. G., Staub, A., Behrens, O. K.: J. Am. Chem. Soc. *79*, 2798 (1957)
103. a) Wünsch, E.: Z. Naturforschg. *22b*, 1269 (1967)
b) Wünsch, E., Zwick, A., Fontana, A.: Chem. Ber. *101*, 326 (1968)
104. Merrifield, R. B., Mojsov, S.: The Chemical Synthesis of Glucagon, in: Glucagon I, Handbook of Experimental Pharmacology, Vol. 66/I (ed. Lefèbvre, P. J.), p. 23, Springer Verlag, Berlin 1983
105. Bodanszky, M., Klausner, Y. S., Said, S. I.: Proc. Natl. Acad. Sci. USA *70*, 382 (1973)
106. a) Tatemoto, K.: Peptides *5*, 151 (1984)
b) Nokihara, K., Yanaihara, C., Iguchi, K., Fukuta, S., Tanaka, M., Mochizuki, T., Tatemoto, K., Lundberg, J. M., Mutt, V., Yanaihara, N.: J. Am. Chem. Soc. *106*, 7909 (1984)
107. Moroder, L., Göhring, W., Thamm, P., Wünsch, E., Tatemoto, K., Mutt, V., Bataille, D.: Z. Naturforschg. *37b*, 772 (1982)
108. Brown, J. C., Dryburgh, J. R.: Can. J. Biochem. *49*, 867 (1971)
109. Gregory, H., Hardy, P. M., Jones, D. S., Kenner, G. W., Sheppard, R. C.: Nature *204*, 931 (1964)
110. a) Gregory, H., Laird, A. H., Morley, J. S., Smith, J. M.: J. Chem. Soc. (C) *1968*, 552
b) Gregory, H., Morley, J. S.: ibid. *1968*, 910
c) Morley, J. S.: ibid. *1969*, 809

d) Briggs, M. T., Morley, J. S.: J. Chem. Soc. Perkin Trans. I *1979*, 2138
111. a) Brown, E., Sheppard, R. C., Williams, B. J.: J. Chem. Soc. Perkin Trans. I *1983*, 1161
b) ibid. *1983*, 75
112. Tam, J. P., Merrifield, R. B.: Int. J. Peptide Protein Res. *26*, 262 (1985)
113. Mutt, V., Jorpes, J. E.: Eur. J. Biochem. *6*, 156 (1968)
114. Wünsch, E., Moroder, L., Wilschowitz, L., Göhring, W., Scharf, R., Gardner, J. D.: Hoppe-Seyler's Z. Physiol. Chem. *362*, 143 (1981)
115. Moroder, L., Wilschowitz, L., Gemeiner, M., Göhring, W., Knof, S., Scharf, R., Thamm, P., Gardner, J. D., Solomon, T. E., Wünsch, E.: Hoppe-Seyler's Z. Physiol. Chem. *362*, 929 (1981)
116. Bodanszky, M., Natarajan, S., Hahne, W., Gardner, J. D.: J. Med. Chem. *20*, 1047 (1977)
117. a) Martinez, J., Bali, J.-P., Magous, R., Laur, J., Lignon, M.-F., Briet, C., Nisato, D., Castro, B.: J. Med. Chem. *28*, 273 (1985)
b) Martinez, J., Bali, J.-P., Rodriguez, M., Castro, B., Magous, R., Laur, J., Lignon, M.-F.: J. Med. Chem. *28*, 1874 (1985)
118. Yajima, H., Futaki, S., Fujii, N., Akaji, K., Funakoshi, S., Sakurai, M., Katakura, S., Inoue, K., Hosotani, R., Tobe, T., Segawa, T., Inoue, A., Tatemoto, K., Mutt, V.: J. Chem. Soc. Chem. Commun. *1985*, 877
119. a) Märki, W., Spiess, J., Taché, Y., Brown, M., Rivier, J. E.: J. Am. Chem. Soc. *103*, 3178 (1981)
b) Minamino, N., Kangawa, K., Hisayuki, M.: Biochem. Biophys. Res. Commun. *119*, 14 (1984)
c) Fujii, N., Futaki, S., Akaji, K., Yajima, H., Inoue, A., Segawa, T.: Chem. Pharm. Bull. *33*, 3731 (1985)
120. Rivier, J. E., Brown, M. R.: Biochemistry *17*, 1766 (1978)
121. Bernardi, L., Bosisio, G., DeCastiglione, R., Goffredo, O.: Experientia *23*, 700 (1967)
122. Cuttitta, F., Carney, D. N., Mulshine, J., Moody, T. W., Fedorko, J., Fischler, A., Minna, J. D.: Nature *316*, 823 (1985)
123. Brown, J. C., Cook, M. A., Dryburgh, J. R.: Can. J. Biochem. *51*, 533 (1973)
124. Wünsch, E., Wendelberger, G., Deimer, K.-H.: Hoppe-Seyler's Z. Physiol. Chem. *357*, 447 (1976)
125. Geiger, R.: Chemiker-Ztg. *100*, 111 (1976)
126. Meienhofer, J., Schnabel, E., Bremer, H., Brinkhoff, O., Zabel, R., Sroka, W., Klostermeyer, H., Brandenburg, D., Okuda, T., Zahn, H.: Z. Naturforschg. *18b*, 1120 (1963)
127. Ullrich, A., Bell, J. R., Chen, E. Y., Herrera, R., Petruzzelli, L. M., Dull, T. J., Gray, A., Coussens, L., Liao, Y.-C., Tsubokawa, M., Mason, A., Seeburg, P. H., Grunfeld, C., Rosen, O. M., Ramachandran, J.: Nature *313*, 756 (1985)
128. Sieber, P., Kamber, B., Hartmann, A., Jöhl, A., Riniker, B., Rittel, W.: Helv. Chim. Acta *57*, 2617 (1974)
129. a) Kovacevic, S., Fayerman, J. T., Miller, J. R., Richardson, M. A.: Brit. Pat. 2 133 798 (1983), E. Lilly
b) Kovacevic, S., Miller, J. R., Hsiung, H. M.: Brit. Pat. 2 133 797 (1983), E. Lilly
130. a) Markussen, J.: BE 887.780 (1981), Novo Industri
b) Markussen, J.: DE 3 229.674 (1982), Novo Industri
131. Tischio, J. P., Patrick, J. E., Weintraub, H. S., Chasin, M., Goldstein, G.: Int. J. Peptide Protein Res. *14*, 479 (1979)
132. Audhya, T., Goldstein, G.: Int. J. Peptide Protein Res. *22*, 187 (1983)
133. König, W., Geiger, R., Obermeier, R., Müllner, H.: DE 3 146 598 (1981), Hoechst AG
134. Goldstein, A. L., Low, T. L. K., McAdoo, M., McClure, J., Thurman, G. B., Rossio, J., Lai, C.-Y., Chang, D., Wang, S.-S., Harvey, C., Ramel, A. H., Meienhofer, J.: Proc. Natl. Acad. Sci. USA *74*, 725 (1977)
135. Wang, S. S., Makofske, R., Bach, A., Merrifield, R. B.: Int. J. Peptide Protein Res. *15*, 1 (1980)
136. Mokotoff, M., Patchornik, A.: Int. J. Peptide Protein Res. *21*, 145 (1983)
137. Wang, S. S., Kulesha, I. D., Winter, D. P.: J. Am. Chem. Soc. *101*, 253 (1979)
138. Felix, A. M., Heimer, E. P., Wang, C.-T., Lambros, T. J., Swistok, J., Roszkowski, M., Ahmad, M., Confalone, D., Scott, J. W., Parker, D., Meienhofer, J., Trzeciak, A., Gillessen, D.: Int. J. Peptide Protein Res. *26*, 130 (1985)
139. Bach, J. F., Dardenne, M., Pleau, J.-M., Rosa, J.: Nature *266*, 55 (1977)
140. a) Abiko, T., Kumikawa, M., Dazai, S., Sekino, H., Higuchi, H.: Chem. Pharm. Bull. *27*, 2207 (1979)
b) Blanot, D., Martinez, J., Anger, G., Bricas, E.: Int. J. Peptide Protein Res. *14*, 41 (1979)

c) Gyotoku, J., Imaizumi, A., Terada, S., Kimoto, E.: Int. J. Peptide Protein Res. *21*, 135 (1983)
141. Najjar, V. A., Nishioka, K.: Nature *228*, 672 (1970)
142. Nishioka, K., Amoscato, A. A., Babcock, G. F., Banks, R. A., Phillips, J. H.: Cancer Investigation *2*, 39 (1984)
143. Wenger, R. M.: Angew. Chem. *97*, 88 (1985)
144. Traber, R., Loosli, H.-R., Hofmann, H., Kuhn, M., von Wartburg, A.: Helv. Chim. Acta *65*, 1655 (1982)
145. Petcher, T. J., Weber, H. P., Rüegger, A.: Helv. Chim. Acta *59*, 1480 (1976)
146. Wenger, R. M.: Helv. Chim. Acta *66*, 2672 (1983)
147. Wenger, R. M.: Helv. Chim. Acta *67*, 502 (1984)
148. Baschang, G., Tarcsay, L., Hartmann, A., Stanek, J.: EP 0 027 258 (1980), Ciba-Geigy
149. Wiesmüller, K.-H., Bessler, W., Jung, G.: Hoppe-Seyler's Z. Physiol. Chem. *364*, 593 (1983)
150. Královec, J.: Drugs of the Future *8*, 615 (1983)
151. Kunz, H., Kauth, H.: Liebigs Ann. Chem. *1983*, 337; ibid. *1983*, 360
152. Krueger, J. M., Pappenheimer, J. R., Karnovsky, M. L.: J. Biol. Chem. *257*, 1664 (1982)
153. Krueger, J. M., Walter, J., Karnovsky, M. L., Chedid, L., Choay, J. P., Lefrancier, P., Lederer, E.: J. Exp. Med. *159*, 68 (1984)
154. Krueger, J. M., Kornovsky, M. L., Martin, S. A., Pappenheimer, J. R., Walter, J., Biemann, K.: J. Biol. Chem. *259*, 12659 (1984)
155. Lernèr, R. A., Green, N., Alexander, H., Liu, F.-T., Sutcliffe, J. G., Shinnick, T. M.: Proc. Natl. Acad. Sci. USA *78*, 3403 (1981)
156. Audibert, F., Jolivet, M., Chedid, L., Alouf, J. E., Boquet, P., Rivaille, P., Siffert, O.: Nature *289*, 593 (1981)
157. Emini, E. A., Jameson, B. A., Wimmer, E.: Nature *304*, 699 (1983)
158. Hemmi, K., Aratani, M., Takeno, H., Okada, S., Miyazaki, Y., Nakaguchi, O., Kitaura, Y., Hashimoto, M.: J. Antibiot. *35*, 1300 (1982)
159. a) Kitaura, Y., Nakaguchi, O., Hemmi, K., Aratani, M., Takeno, H., Okada, S., Tanaka, H., Hashimoto, M.: EP 0 027 260 (1980), Fujisawa Pharmaceutical
b) Kitaura, Y., Nakaguchi, O., Hemmi, K., Yonishi, S., Takeno, H., Okada, S., Hashimoto, M.: EP 0 057 419 (1982), Fujisawa Pharmaceutical
160. Izumi, S., Nakahara, K., Gotoh, T., Hashimoto, S., Kino, T., Okuhara, M., Aoki, H., Imanaka, H.: J. Antibiot. *36*, 566 (1983)
161. Mine, Y., Yokota, Y., Wakai, Y., Fukuda, S., Nishida, M., Goto, S., Kuwahara, S.: J. Antibiot. *36*, 1045 (1983)
162. Mine, Y., Watanabe, Y., Tawara, S., Yokota, Y., Nishida, M., Goto, S., Kuwahara, S.: J. Antibiot. *36*, 1059 (1983)
163. Closse, A., Huguenin, R.: Helv. Chim. Acta *57*, 553 (1974)
164. a) Rich, D. H., Jasensky, R. D.: J. Am. Chem. Soc. 101, 5413 (1979)
b) Pastuszak, J., Gardner, J. H., Singh, J., Rich, D. H.: J. Org. Chem. *47*, 2982 (1982)
c) Kawai, M., Jasensky, R. D., Rich, D. H.: J. Am. Chem. Soc. *105*, 4456 (1983)
165. Schmidt, U., Beuttler, T., Lieberknecht, A., Griesser, H., Tetrahedron Lett. *24*, 3573 (1983)
166. Chakravarty, P. K., Carl, P. L., Weber, M. J., Katzenellenbogen, J. A.: J. Med. Chem. *26*, 633 (1983)
167. Umezawa, H., Aoyagi, T., Shirai, T., Nishizawa, R., Suzuki, M., Saino, T.: DOS 2 947 140 (1979), Nippon Kayaku
168. Drugs of the Future *6*, 604 (1981)
169. Cushman, D. W., Ondetti, M. A.: Prog. Med. Chem. *17*, 41 (1980)
170. Ondetti, M. A., Cushman, D. W.: J. Med. Chem. *24*, 355 (1981)
171. Khosla, M. C., Hall, M. M., Smeby, R. R., Bumpus, F. M.: J. Med. Chem. *17*, 431 (1974)
172. Drugs of the Future *6*, 542 (1981)
173. Bumpus, F. M., Khosla, M. C., Smeby, R. R.: DOS 2 520 106 (1975), Abbott Laboratories
174. Drugs of the Future *6*, 634 (1981)
175. Drugs of the Future *6*, 527 (1981)
176. Geiger, R.: Arzneim.-Forsch./Drug Res. *34* (II), 1386 (1984)
177. Patchett, A. A., Harris, E., Tristam, E. W., Wyvratt, M. J., Wu, M. T., Taub, D., Peterson, E. R., Ikeler, T. J., ten Broeke, J., Payne, L. G., Ondeyka, D. L., Thorsett, E. D., Greenlee, W. J., Lohr, N. S., Hoffsommer, R. D., Joshua, H., Ruyle, W. V., Rothrock, J. W., Aster, S. D., May-

cock, A. L., Robinson, F. M., Hirschmann, R., Sweet, C. S., Ulm, E. H., Gross, D. M., Vassil, T. C., Stone, C. A.: Nature *288*, 280 (1980)
178. Teetz, V., Geiger, R., Henning, R., Urbach, H.: Arzneim.-Forsch./Drug Res. *34* (II) 1399 (1984)
179. Greenlee, W. J., Allibone, P. L., Perlow, D. S., Patchett, A. A., Ulm, E. H., Vassil, T. C.: J. Med. Chem. *28*, 434 (1985)
180. Gruenfeld, N., Stanton, J. L., Yuan, A. M., Ebetino, F. H., Browne, L. J., Gude, C., Huebner, C. F.: J. Med. Chem. *26*, 1277 (1983)
181. Ksander, G. M., Yan, A. M., Diefenbacher, C. G., Stanton, J. L.: J. Med. Chem. *28*, 1606 (1985)
182. Attwood, M. R., Hassall, C. H., Lambert, R. W., Lawton, G. R.: DOS 3 317 290 (1983), Hoffmann-La Roche, Schweiz
183. Skeggs, L. T., Lentz, K. E., Kahn, J. R., Hochstrasser, H.: J. Exp. Med. *128*, 13 (1968)
184. a) Szelke, M., Leckie, B., Hallett, A., Jones, D. M., Sueiras, J., Atrash, B., Lever, A. F.: Nature *299*, 555 (1982)
b) Leckie, B., Szelke, M., Hallet, A., Hughes, M., Lever, A. F., McIntyre, G., Morton, J. J., Tree, M.: Clin. Exp. Hypertens. *A 5*, 1221 (1983)
c) Haber, E.: Clin. Exp. Hypertens. *A 5*, 1193 (1983)
185. Boger, J., Lohr, N. S., Ulm, E. H., Poe, M., Blaine, E. H., Fanelli, G. M., Lin, T.-Y., Payne, L. S., Schorn, T. W., La Mont, B. I., Vassil, T. C., Stabilito, I. I., Veber, D. F., Rich, D. H., Bopari, A. S.: Nature *303*, 81 (1983)
186. Rittle, K. E., Homnick, C. F., Ponticello, G. S., Evans, B.-E.: J. Org. Chem. *47*, 3016 (1982)
187. a) Veber, D. F., Boger, J.: EP 0 077 028 (1982), Merck, USA
b) Veber, D. F., Rich, D. H.: EP 0 077 029 (1982), Merck, USA
188. Boger, J.: Proc. of the 8th Am. Peptide Symp., p. 569, Tucson 1983
189. a) Rich, D. H., Salituro, F. G., Holladay, M. W.: Proc. of the 8th Am. Peptide Symp., p. 139, 511, Tucson 1983
b) Rich, D. H., Sun, E. T. O.: J. Med. Chem. *23*, 27 (1980)
c) Rich, D. H.: ibid. *28*, 263 (1985)
190. a) Bock, M. G., Boger, J. S., Brady, S. F. Veber, D. F.: EP 0 157 409 (1985), Merck, USA
b) Boger, J. S.: EP 0 156 318 (1985), EP 0 156 319 (1985) and EP 0 156 320 (1985), Merck, USA
c) Bock, M. G., Dipardo, R. M., Boger, J. S.: EP 0 156 321 (1985), Merck, USA
d) Boger, J. S., Bock, M. G.: EP 0 156 322 (1985), Merck, USA
e) Boger, J., Payne, L. S., Perlow, D. S., Lohr, N. S., Poe, M., Blaine, E. H., Ulm, E. H., Schorn, T. W., LaMont, B. I., Lin, T.-Y., Kawai, M., Rich, D. H., Veber, D. F.: J. Med. Chem. *28*, 1779 (1985)
191. Boissonas, R. A., Guttmann, St., Jaquenoud, P.-A.: Helv. Chim. Acta *43*, 1349 (1960); ibid. *43*, 1481 (1960)
192. a) Bertaccini, G.: Pharmacol. Rev. *28*, 127 (1976)
b) Miller, R. J.: J. Med. Chem. *27*, 1239 (1984)
193. Sandrin, E., Boissonas, R. A.: Experientia *18*, 59 (1962)
194. Lübke, K., Schröder, E., Schmiechen, R., Gibian, H.: Liebigs Ann. Chem. *679*, 195 (1964)
195. a) Chillemi, F., Bernardi, L., Bosisio, G.: Gazz. Chim. Ital. *94*, 891 (1964)
b) Bernardi, L., Bosisio, G., Chillemi, F., DeCaro, G., DeCastiglione, R., Erspamer, V., Glaesser, A., Goffredo, O.: Experientia *20*, 306 (1964)
c) Buck, S. H., Burcher, E.: Trends in Pharmacological Sciences *7*, 65 (1986)
196. Voelter, W., Zech, K., Jung, G., Sewing, K.-F.: Tetrahedron *28*, 2649 (1972)
197. Chillemi, F.: Gazz. Chim. Ital. *95*, 402 (1965)
198. Voelter, W., Zech, K., Jung, G., Sewing, K.-F.: Tetrahedron *28*, 5963 (1972)
199. Anastasi, A.: Experientia *33*, 857 (1977)
200. Bernardi, L., Bosisio, G., DeCastiglione, R.: Experientia *23*, 700 (1967)
201. Zetler, G.: DE 3 138 233 (1981), Farmitalia-Carlo Erba
202. Flynn, T. G., de Bold, M. L., de Bold, A. J.: Biochem. Biophys. Res. Commun. *117*, 859 (1983)
203. a) Kangawa, K., Matsuo H., Biochem. Biophys. Res. Commun. *118*, 131 (1984)
b) Kangawa, K., Fukada, A., Matsuo, H.: Nature *313*, 397 (1985)
204. Kangawa, K., Tawaragi, Y., Oikawa, S., Mizuno, A., Sakuragawa, Y., Nakazato, H., Fukuda, A., Minamino, N., Matsuo, H.: Nature *312*, 152 (1984)
205. a) Seidah, N. G., Lazure, C., Chrétien, M., Thibault, G., Gracia, R., Cantin, M., Genest, J.,

Nutt, R. F., Brady, S. F., Lyle, T. A., Paleveda, W. J., Colton, C. D., Ciccarone, T. M., Veber, D. F.: Proc. Natl. Acad. Sci. USA *81*, 2640 (1984)
b) Nutt, R. F., Brady, S. F., Lyle, T. A., Dylion-Colton, C., Paleveda, W. J., Ciccarone, T. M., Blaine, E. H., Winquist, R. J., Bennet, C. D., Hirschmann, R., Veber, D. F.: Proc. of 18th Eur. Peptide Symp., p. 513, Djurönäset 1984, Sweden
206. a) Currie, M. G., Geller, D. M., Cole, B. R., Siegel, N. R., Fok, K. F., Adams, S. P., Eubanks, S. R., Galluppi, G. R., Needleman, P.: Science *1983*, 67
b) Needleman, P.: US-Pat. 4 496 544 and 4 508 712 (1985), Washington University, St. Louis
207. Fok, K. F., Tjoeng, F. S., Houbion, J. A., Spear, K. L., Nugent, S. T., Eubanks, S. R., Zupec, M. E., Olins, G., Blehm, D. J., Adams, S. P., Wakitani, K., Needleman, P.: Proc, of the 9th Am. Peptide Symp., p. 953, Toronto 1985
208. Riniker, B., Brugger, M., Kamber, B., Sieber, P., Rittel, W.: Helv. Chim. Acta *52*, 1073 (1969)
209. Sakakibara, S., Noda, T., Morikawa, T., Munekata, E., Kimura, T., Nakagawa, Y.: DOS 2 616 399 (1976), Toyo Jozo K. K.
210. Orlowski, R. C., Stahl, G. L., Colescott, R. L.: US-Pat. 4 528 132 (1984), Armour Pharmaceutical Co., USA
211. Izumiya, N., Kato, T., Aoyagi, H., Waki, M., Kondo, M.: Synthetic Aspects of Biologically Active Cyclic Peptides, J. Wiley & Sons, New York 1979
212. Hassall, C. H., Johnson, W. H., Theobald, C. J.: J. Chem. Soc. Perkin Trans. I. *1979*, 1451
213. Kupryszewski, G.: Synthetic Oligopeptide Derivatives with Antibacterial Activity, in: Perspectives in Peptide Chemistry (ed. Eberle, A., Geiger, R., Wieland, T.), p. 307, S. Karger, Basel 1981
214. Alewood, P. F., Perich, J. W., Johns, R. B.: Synthetic Commun. *12*, 821 (1982)
215. Maier, L.: Phosphorus and Sulfur, Vol. 14, 295 (1983)
216. a) Atherton, R. F., Hall, J. M., Hassall, C. H., Lambert, R. W., Ringrose, P. S.: DOS 2 730 524 (1977); DOS 2 730 549 (1977), Hoffmann-La Roche, Switzerland
b) Atherton, F. R., Hassall, C. H., Lambert, R. W.: J. Med. Chem. *29*, 29 (1986)
217. Meienhofer, J., Patel, R. P.: Int. J. Peptide Protein Res. *3*, 347 (1971)
218. a) Benz, G.: Liebigs Ann. Chem. *1984*, 1399
b) Metzger, K. G., Pfitzner, J., Schmidt, D., Weyland, H., Benz, G., Schröder, T.: DOS 3 102 137 (1981); DOS 3 102 136 (1981), BAYER AG
219. Gross, E., Morell, J. L.: J. Am. Chem. Soc. *92*, 2919 (1970), ibid. *93*, 4634 (1971)
220. Wakamiya, T., Shimbo, K., Sano, A., Fukase, K., Shiba, T.: Bull. Chem. Soc. Jpn. *56*, 2044 (1983)
221. Hagenmaier, H., Keckeisen, A., Zähner, H., König, W. A.: Liebigs Ann. Chem. *1979*, 1494
222. a) Hass, W., König, W. A.: Liebigs Ann. Chem. *1982*, 1615
b) Rathmann, R., König, W. A., Schmalle, H., Carlsson, G., Bosch, R., Hagenmaier, H., Winter, W.: Liebigs Ann. Chem. *1984*, 1216
223. a) König, W. A., Hass, W., Dehler, W., Fiedler, H.-P., Zähner, H.: Liebigs Ann. Chem. *1980*, 622
b) Zimmermann, G., Hass, W., Faasch, H., Schmalle, H., König, W. A.: Liebigs Ann. Chem. *1985*, 2165.
224. Merrifield, R. B., Vizoli, L. D., Boman, H. G.: Biochemistry *21*, 5020 (1982)
225. Andreau, D., Merrifield, R. B., Steiner, H., Boman, H. G.: Proc. of the 18th Eur. Peptide Symp., p. 541, Djurönäset 1984, Sweden
226. Payne, J. W., Jakes, R., Hartley, B. S.: Biochem. J. *117*, 757 (1970)
227. a) Schmitt, H., Jung, G.: Liebigs Ann. Chem. *1985*, 321
b) Jung, G., Brückner, H., Schmitt, H.: Properties of the Membrane Modifying Polypeptide Antibiotics Alamethicin and Trichotoxin A-40, in: Structure and Activity of Natural Peptides, (ed. Voelter, W., Weitzel, G.), p. 75, Walter de Gruyter, Berlin 1981
228. a) Bartmann, W., Beck, G., Angew. Chem. *94*, 767 (1982)
b) Samuelsson, B.: Science *220*, 568 (1983)
229. Corey, E. J., Clark, D. A., Goto, G., Marfat, A., Mioskowski, C.: J. Am. Chem. Soc. *102*, 1436 (1980)
230. a) Gleason, J. G.: Abstracts of the 2nd SCI-RSC Medicinal Chemistry Symposium, Cambridge 1983
b) Saksena, A. K., Green, M. J., Mangiaracina, P., Wong, J. K., Kreutner, W., Gulbenkian, A. R.: Tetrahedron Lett. *26*, 6423 (1985)

231. Shiba, T.: Peptide Chemistry 1977, Proc. of the 15th Symp. on Pept. Chem., Osaka 1977
232. Bayer, E., Gugel, K. H., Hägele, K., Hagenmaier, H., Jessipow, S., König, W. A., Zähner, H.: Helv. Chim. Acta *55*, 224 (1972)
233. a) Wieland, T.: Chemie in unserer Zeit *13*, 56 (1979)
b) Witkop, B.: Naturwissenschaftliche Rundschau *36*, 261 (1983)
234. Wieland, T., Faesel, J., Konz, W.: Liebigs Ann. Chem. *722*, 197
235. a) Bodanszky, M., Stahl, G. L.: Proc. Natl. Acad. Sci. USA *71*, 2791 (1974)
b) Bodanszky, M., Bednarek, M. A., Yiotakis, A. E., Curtis, R. W.: Int. J. Peptide Protein Res. *20*, 16 (1982)
236. Williams, D. H., Santikarn, S., DeAngelis, F., Smith, R. J., Reid, D. G., Oelrichs, P. B., McLeod, J. K.: J. Chem. Soc. Perkin Trans. I, *1983*, 1869
237. a) Mazur, R. H., Schlatter, J. M., Goldcamp, A. H.: J. Am. Chem. Soc. *91*, 2684 (1969)
b) Mazur, R. H., Reuter, J. A., Swiatek, K. A., Schlatter, J. M.: J. Med. Chem. *16*, 1284 (1973)
238. a) Otagiri, K., Miyake, I., Ishibashi, N., Fukui, H., Kanehisa, H., Okai, H.: Bull. Chem. Soc. Jpn. *56*, 1116 (1983)
b) Otagiri, K., Shigenaga, T., Kanehisa, H., Okai, H.: ibid. *57*, 90 (1984)
239. Fuller, W. D., Goodman, M., Verlander, M. S.: J. Am. Chem. Soc. *107*, 5821 (1985)

Author Index Volumes 101–136

Contents of Vols. 50–100 see Vol. 100
Author and Subject Index Vols. 26–50 see Vol. 50

The volume numbers are printed in italics

MIX
Papier | Fördert gute Waldnutzung
FSC® C083411

Zeitfracht Medien GmbH
Ferdinand-Jühlke-Straße 7
99095 Erfurt, Deutschland
produktsicherheit@kolibri360.de